MAN VS BIG DATA

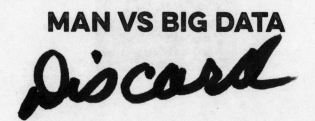

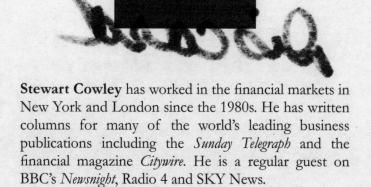

Stewart Cowley has worked in the financial markets in New York and London since the 1980s. He has written columns for many of the world's leading business publications including the *Sunday Telegraph* and the financial magazine *Citywire*. He is a regular guest on BBC's *Newsnight*, Radio 4 and SKY News.

MAN VS
BIG DATA

EVERYDAY DATA
EXPLAINED

Stewart Cowley

Illustrated by Joe Lyward

Aurum
Press

Brimming with creative inspiration, how-to projects and useful information to enrich your everyday life, Quarto Knows is a favourite destination for those pursuing their interests and passions. Visit our site and dig deeper with our books into your area of interest: Quarto Creates, Quarto Cooks, Quarto Homes, Quarto Lives, Quarto Drives, Quarto Explores, Quarto Gifts, or Quarto Kids.

First published in 2017 by Aurum Press
an imprint of The Quarto Group
The Old Brewery, 6 Blundell Street
London N7 9BH
United Kingdom

www.QuartoKnows.com

A catalogue record for this book is available from the British Library.

ISBN 978 1 78131 669 6
Ebook ISBN 978 1 78131 756 3

10 9 8 7 6 5 4 3 2 1
2021 2020 2019 2018 2017

Illustrations and design by Joe Lyward

Printed by CPI Group (UK) ltd, Croydon, CR0 4YY

MIX
Paper from
responsible sources
FSC® C020471
www.fsc.org

Contents

Introduction

The parents of six-year-old Brooke Neitzel stared down at their daughter, her hands clasped together as though in prayer. Her hope was that the $160 KidKraft Girl's Uptown Dollhouse with Furniture and the four pounds of Danish butter cookies that had just arrived were here to stay.

Brooke, in a moment of profound desire, had expressed her love for the items within earshot of Alexa – the voice-activated virtual assistant that can turn your whims into a brown cardboard package delivered to your door from Amazon. All Brooke had said to Alexa was, 'Can you play dollhouse with me and get me a dollhouse?' After confirming the order, Brooke skipped away and began her vigil by the front door. When the incident was reported on a local TV station, six other Alexas heard it and promptly tried to order the KidKraft Girl's Uptown Dollhouse for their respective owners.

The Neitzels' story has the hand of Big Data all over it. Alexa uses voice recognition-powered artificial intelligence software created by analyzing thousands of voice samples to create a virtual assistant that responds in less a second. Amazon then processes any order and the transaction is added to the data profile of her parents. The KidKraft business intelligence unit may notice the slight demand increase after the TV report and raise the price of the Girl's Uptown Dollhouse by 25¢. The KidKraft transaction data can now be aggregated and sold on to third parties to help them guide their sales to the Neitzels. Pretty soon, advertisements for similar and related products start popping up in their YouTube accounts, Google sidebar, Facebook pages and email accounts. Manufacturers of similar toys are alerted to the success of the KidKraft campaign and book advertising spaces on the local television stations. The payment for the Girl's Uptown Dollhouse will be processed by the Neitzels' credit card company, which has software detecting where Brooke is in her life cycle, prompting a financial adviser to begin sending them suggestions for financial products to help save for Brooke's college fund. After the AI-powered logistics at Amazon have packaged Brooke's dollhouse (and ordered a replacement for stock control) the box is passed to FedEx for delivery. FedEx uses a program utilizing Big Data which plots the most efficient route for drivers (mainly by avoiding turning left). Music streaming service Spotify is notified by the affiliates it shares data with about the purchase and adds 'Castles in the Air' by Don McLean to Mr. Neitzel's 'Discover Weekly' list. Finally, the package arrives at the Neitzels' door – to Brooke's delight and her parents' bewilderment.

This is only a sample of how some of the processes falling under the umbrella term 'Big Data' have penetrated our lives, sometimes overtly and sometimes imperceptibly. Whether we know it or not, or if we like it or not, we now leave behind us a trail of data wherever we go. Amazon's Alexa listens to all speech by default. Buying something online with a credit card generates data stored as a transaction history at a retailer and at the finance company. Switching on your mobile phone with the GPS tracker activated creates data on where you have been, which can be collected by the network provider or government. Switching on your washing machine generates data about what washing cycles you use and how frequently you use it, which some privacy agreements allow the machine to send back to the manufacturers if it is connected to the Internet via your home wifi. You simply can't help generating data any more – and it's all being recorded, stored, analyzed and, in many cases, the conclusions acted upon in startling ways.

Needless to say, there is now a lot at stake when it comes to data; it has become a valuable commodity in itself. Companies trawl through your trail to detect your behaviour and its subtle changes to position themselves better and sell you new products. Employers are monitoring your effectiveness and work habits to find out if you are a good employee and what they need you to do to increase profits. Governments monitor Internet and social media traffic to help prevent crime and increase national security. The Big Data era has been described as the 'Fourth Industrial Revolution', coming as it does after mechanization (the Industrial Revolution in the late eighteenth century), technological breakthroughs in mass communication, transport and industrial production (the Second Industrial Revolution of the late nineteenth century) and the combination of new communications and manufacturing technology with renewable energy sources (the Third Industrial Revolution of the early twenty-first century). The Big Data era not only ushers in opportunities for consumers and producers, but also shapes our politics, science and our relationship with work. It may even start directing who we marry and who we choose to have children with.

Big Data offers a tantalizing glimpse of the future, where real-time trends are spotted and acted upon to avoid disasters, provide much-needed services (and even ones you didn't know you needed) and aid global economic growth. For the twenty-first century Big Data is, potentially, what water and steam power was to the first Industrial Revolution. Its effects will be felt for generations to come.

To explore Big Data, you have to experience it first-hand. In order to write this book I have:

- Joined a dating website
- Put my marriage through a compatibility algorithm
- Opened an online gambling account
- Experimented on myself to find out if I could run a marathon
- Read all the privacy agreements I am signed up to
- Had my genome analyzed

Some of what I experienced came as a surprise to me – and some of it might come of a surprise to you. If you don't know anything about the subject of Big Data, now is the time to find out. There are urgent questions to ask. For all its benefits, should we be wary of Big Data? Are we being snooped on, directed and controlled like never before? Should we care? What does it mean for the distribution of wealth in the world? Can I beat the bookies using Big Data? How do I find a perfect mate using Big Data? This book won't give you all the answers, but it will, hopefully, set you on the road to a journey into something that is already having a huge impact on all our lives.

Big Data vs Big

Is there enough room in the universe to store all our selfies?

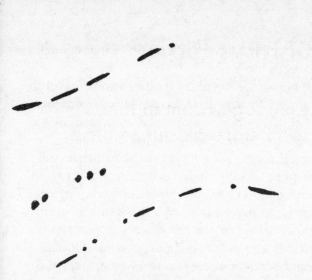

In 1844 Samuel F.B. Morse sent the first official telegram. The slightly ominous biblical phrase 'What hath God wrought', chosen by Annie Ellsworth, the young daughter of the Patent Commissioner who helped to fund the invention, heralded a sea change in our ability to communicate information speedily; it was the beginning of the digital age. Words were now being converted into dots and dashes and transmitted.

By 1863, Edward A. Calahan had converted the telegraph into a machine that could reliably transmit and print out stock prices on long rolls of paper. The machine, called a ticker tape because of the clacking sound it made, became so ubiquitous it gave birth to the ticker tape parade: the tradition of cascading used rolls of paper over heroes and dignitaries as they paraded through American cities in open-top cars. Ever since 1863, all we have done is refine the methods of doing the same thing; what was once dots and dashes converted to text on paper is now electrons in a solid-state hard drive converted to legible text on a screen. Once it got going, we never looked back.

Simple messaging has today evolved into telling our personal stories in real time: camera-based recording creatively edited on a smartphone and distributed through various online platforms has become more than commonplace – it's everywhere. Meanwhile, our digital footprint stamps itself on the world through every interaction we have. Our televisions, fitness devices, games consoles, cars, smartphones and every app we glance at – even our movements at work – are all pouring data into private and public storage devices, where it can be filtered and analyzed by companies and governments to be used in ways that were unimaginable even a decade ago.

We are now storing data at an unprecedented rate, but no matter how commonplace it is to record ourselves or have others record us, there is a lot of confusion about what 'data' is and how it is stored. When discussing Big Data, experts throw around terms like 'bits', 'bytes', 'megabytes', 'terabytes', or even something called a 'yotta', with alarming casualness. Why are these mysterious words so important to understanding the era of Big Data?

We'll start by looking at how text is stored and see how the size of a computer file grows when we introduce pictures and sound. Then, as information is being stored at an ever-increasing rate, we'll examine whether we could ever reach the limit of our ability to store data in the near future.

Of bits and bytes...

To understand why Big Data got so big, we need to start at Morse's telegraph. It was based on the relay switch, where an operator pressed a button connecting two pieces of metal together, allowing an electrical current to pass through a wire wrapped around a soft iron core, turning it into a magnet. The magnet attracted a piece of steel which, in turn, clicked into place completing a circuit. The current then passed down a wire where, at the other end, a similar device was triggered and a rather pleasing 'click' was heard.

Morse Code relied on the length of the circuit to produce combinations of dots and dashes to represent letters, but it didn't change the fact that it was either 'off' or 'on'. This was a big breakthrough: the two off and on states could also be described as 0 or 1 – the basis of the binary system.

The first computers were essentially a giant collection of switches, each one triggering the next in logical order depending on what the initial input was. Think of it like a floor full of mousetraps – once you start off the first, it triggers a reaction in the next and so on. Using switches arranged in the correct sequence – Boolean logic, as it is known – allows you to ask questions like, 'What is the sum of 4 + 1?' As the cascade of switches ripples through the system, it eventually outputs a combination of switches indicating the answer '5'. Eventually, the relay switch was replaced by tiny circuits printed into semiconducting materials and the silicon chip was born. This process of miniaturization brought speed and power to computing, but the basic principles have never really changed.

Computing answers is fine if you only want to ask a question once. What about if you want to store the answer? In that case, the computer needs to be able to interpret it as a piece of binary data. To understand how computers use the binary data system to store data, let's learn to count in binary like a computer does. Forget your usual counting, which moves up a digit each time you ratchet up another ten units (known as base-10). In binary, you step up a digit when the previous number doubles. So instead of the series 1, 10, 100, 1000 etc. in binary the series is 1, 2, 4, 8, 16 etc.

To visualise this think of a row of light bulbs. If the light bulb is off it is represented by a 0 and if it is on it is represented by a 1. The light bulbs are arranged in the same assending order as the binary system and are switched on when we input a base-10 number. By looking at which bulbs are on and off, we can read off what the binary equivalent is. So, the number 5 in base-2 looks like this:

128   64   32   16   8   4   2   1

It consists of 1×4 and 0×2 and 1×1. The binary equivalent number is 101. Or if you wanted to represent 42 it would look like this:

128   64   32   16   8   4   2   1

Which consists of 1×32, 0×16, 1×8, 0×4, 1×1 and 0×1 making the binary equivalent number 101010.

The combinations of zeros and ones can be used to create logical functions or represent a letter or symbol in the American Standard Code for Information Interchange (ASCII) format. For instance, the letter 's' has an ASCII code of 115. In binary, the number 115 is 01110011. This is what is recorded in a computer file rather than an 's'. Notice that no matter how big the number, it always has eight digits or 'bits' associated with it – this is important, as we'll see later.

If we were to write out my name, inside a computer file it would actually look like this:

Symbol	s	t	e	w	a	r	t
ASCII Code	115	116	101	119	97	114	116
Binary	01110011	01110100	01100101	01110111	01100001	etc.	

The binary, 8-bit series runs from 00000000 to 11111111, or, in other words, 0 to 255 in base-10, giving 256 characters. There's no particular reason why eight bits are used; it's just something everyone settled on as 255 seems to cover enough characters embrace many situations. To find out how many characters there are in a file you divide the number bits by eight: eight bits make a byte.

Enough to strike Tera into your heart...

To illustrate how this works in practice and what it does to file sizes, let's try an experiement. Open a text application on your computer (Notepad if using Windows or TextEdit for Mac) and type the following, rather immodest, piece of text and save it as plain text to your desktop (include the full stop):

stewart cowley is great.

Now have a look at the size of the file. If you have done this correctly and you have used a pure text editor, then the size of the file is 24 bytes. Each of the 24 characters (including the three spaces) corresponds to one byte.

You are probably familiar with words like 'kilo', 'mega' and 'giga' and you may even be numerate enough to associate these words with 1,000, 1 million and 1 billion respectively. These can also be written as 103, 106 and 109. However, in the binary system we are working in a base of two and when you start raising it to the power it starts to deviate away from our base-10 numbers (this is important, as we will see later). So a 'kilo' in base-10 would be associated in most people's minds as 1,000 – but in the world of data it turns into 1,024, which is actually 210 (the closest we can get in nice round numbers, otherwise it would be 29.9657843

= 1,000 which isn't very satisfactory). Stepping through two to the power of increasing tens you get the sequence:

Name	Power	Number of BYTES	Base 10 Equivalent[1]
Kilo (KB)	2^{10}	1,024	10^3 A thousand
Mega (M)	2^{20}	1,048,576	10^6 A million
Giga (G)	2^{30}	1,073,741,824	10^9 A billion
Tera (T)	2^{40}	1,099,511,627,776	10^{12} A trillion
Peta (P)	2^{50}	1,125,899,906,842,624	10^{15} Quadrillion
Exa (E)	2^{60}	1,152,921,504,606,846,976	10^{18} Sextillion
Zetta (Z)	2^{70}	1,180,591,620,717,411,303,424	10^{21} Septillion
Yotta (Y)	2^{80}	1,208,925,819,614,629,174,706,176	10^{24} Octillion

Notice how, as we go higher and higher up the order, the base-10 numbers deviate from the binary equivalent until we get to a yotta, where there is a 20.89 per cent difference between the two numbers. Big Data is bigger than you think.

The binary definitions of kilo, mega, and so on, are used to convert the large numbers of bytes into more manageable numbers with names. For instance, if you had a file with 6,144 characters it wouldn't be 6.144KB, it would 6KB (6,144 divided by 1,024).

If you think that these numbers are otherworldly, you are only half right: it's estimated that there are about 1,024 stars in the known, observable universe. In other

words, by the time we reach our first yotta of data in the Big Data era, there will be 20 per cent more data bytes on Earth than there are stars in the known universe. This still doesn't explain how data got 'big': even an average 60,000-word novel, allowing for spaces, isn't going to be more than 250,000 to 300,000 bytes. So where is all the other stuff coming from?

Two hundred and fifty-six shades of grey...

12MP, 4000x3000 pixels

iPhone 6 Plus, 5S, 5, 4S: 8MP, 3264x2448 pixels

iPhone 4, iPad 3, iPod Touch 2012: 5MP, 2592x1936 pixels

iPhone 3GS: 3.2MP, 2048x1536 pixels

iPhone 5 Screen: 640x1136

iPhone 2G, 3G: 2MP, 1600x1200 pixels

Instagram-friendly: 640x640

320x480

Compared to the time when the Bayeux Tapestry was regarded with the same awe as the first *Toy Story* movie, we now live in an age of casually dismissive documentation. We simply can't help ourselves when it comes to recording every twitch of our daily lives, while anything we touch using our computers, domestic applicances or smartphones generates data. Nothing illustrates this better than our relationship with vision and sound. Take, for instance, the development of images taken on Apple's iPhone.

All digital images – every 'selfie', party pic or picture of your dinner – are split into tiny squares called pixels. It's the smallest controllable element of an image. The pixels are arranged in a neat grid with a width and a height. We used to thrill at the idea of images 320 × 480 pixels (153,600 pixels) in size, but now it isn't unusual to find images on a standard mobile device measuring up to 4,000 × 3,000 pixels (12,000,0000 pixels).

To produce an image, each one of these dots needs a colour assigned to it. In the simplest format they could be just black and white squares representing two colours stored in the picture file. The image is said to have a bit depth of 1 because we only need one digit (either 1 or 0) to represent the colours. So a 320 × 480 pixel picture would have a file size of

$$320 \times 480 \times 1 = 153{,}600 \text{ bits or } 153{,}600/8 = 19{,}200 \text{ bytes}$$
$$= 19{,}200/1{,}024 \approx 18.8\text{KB}$$

In terms of file size, 18.8KB is trivial and will produce the kind of crude image a housefly might recognise, but would look pretty poor to a human being. So you need a lot of shades in between black and white to create a decent image. Using the binary system (like we did with letters) to describe each step between white and black creates a grey scale of eight bits per pixel (giving 28 or 256 possible shades between pure black and pure white).

This desire for detail has consequences in the data world. Now our 320 × 480 image has a file size of

$$320 \times 480 \times 8 = 1,228,800 \text{ bits or}$$

$$1,228,800/8 = 153,600 \text{ bytes or, using our conversion,}$$

$$153,600/1,024 = 150\text{KB}$$

A 150KB image isn't big in the data world either, but you'll notice just by this small change we have achieved an eight-fold increase in the size of the image file.

Now let's expand the image size, and watch the file size grow – and remember, this is just for a black and white picture. Do the calculations yourself and, by the time you get to the iPhone 6 with its $3,264 \times 2,448$ pixel images, each image file is nearly eight megabytes (eight million bytes) in size. That's a big leap up from our original rather piffling 19,200 bytes.

Now let's start adding colour. You can create every colour in the rainbow by using a combination of red, green and blue. If each of these colours had its own 8-bit channel you would have 24 bits of information per pixel and 224 combinations or 16,777,216 possible colours, which should be enough for anybody.

Do the calculations again and what you will find is even the smallest image has become 400KB and by the time you get to the iPhone 5 and 6, the image size is now just under an astonishing 23MB for one image. According to Mary Meeker's 2016 Internet Trends Report, the combined number of images per day being shared via platforms such as Facebook, Snapchat, Instagram and WhatsApp has risen from virtually nothing in 2008 to over 3 billion today. That's a trillion photographs a year.

Clearly this is not a sustainable situation, so data scientists play a trick on us: they lose all the stuff our brains don't need while still giving us enough information to construct a high-quality image. It's a mathematical process called quantization, where compressing the image (by as much as ten times) creates a trade-off between image quality and file size. This is where the JPEG, PNG and TIFF at the end of image files come from. Each of these formats is a method of making files smaller while keeping a semblance of quality.

The fastest-growing part of our recorded lives is through video. In the two years between 2014 and 2016, the number of user-shared video views on Snapchat and Facebook grew from just one view per day to ten. Let's now add in the dimension of time to our images and see what it does to the amount of data we are storing.

We are all pretty much familiar with flip-book animations that create the illusion of movement through time, with the need for a separate image to be moved on slightly and played back rapidly to create a sense of motion. Video images work in the same way, but they require anything between 25 (PAL) and 30 (NTSC) frames per second. Let's examine the enormity of the files created, even for a modest 640 × 480 pixel screen with a 24-bit colour system.

Each frame = 640 × 480 × 24 = 7,372,800 bits per frame

Bytes per frame = 7,372,800/8 = 921,600

In Megabytes = 921,600/1,048,576 = 0.88MB

At 25 frames per second = 25 × 0.88 =21.97MB per second

In one minute = 22.97 × 60 = 1,318.36Mb = 1.29GB

In one hour = 60 × 1.29Gb = 77.25GB

Again, this is a simply enormous amount of data. To cope with it and to prevent your hard drive or smartphone becoming clogged up within a few hours, video images undergo a process of compression, just like single frames, to reduce the actual file sizes.

We haven't yet even added sound. Digitised sound uses similar principles to images to record soundwaves as a series of zeros and ones: an analogue recording is sampled and each sample given a sound 'colour' on a 16-bit spectrum (8-bits per each channel for stereo). For instance, a standard 118mm compact disc can hold 700MB of data, the equivalent of about 80 minutes of sound. Audio files can be in different formats (such as WAV or MP3), and to get there they go through a process of compression to reduce their file size. Add one hour's worth of sound to a movie (747MB of data converts into 74 minutes of uncompressed sound) and now you start to understand why you could get to an uncompressed file size of some 137GB of data just for a single hour of entertainment. Most movies last about two hours.

To give the impression that the Big Data revolution has only been about the spectacular growth of images would be wrong. The other rich source of data is you – someone is watching the watchers...

Every breath you take...

We've collected data about each other for as long as we've been able to record it. The 1970s were altogether simpler and more direct times: rummaging through trashcans, as pioneered by William L. Rathje at the University of Arizona, showed that even recent artefacts could tell you a lot about the people who used and discarded them. 'Garbology' became a valuable method of defining social trends.

In many respects we are continuing the rummaging tradition today, but instead of going through what you leave behind, data scientists are doing it in real time whilst they recite the mantra of the four 'V's':

Velocity
Volume
Variety
Veracity

This basically means that a lot of data collected very quickly from lots of places with no apparent connection to each other is the most valuable. The best source of this disconnected vision of your behaviour is your personal computer, smartphone and domestic appliances.

If you are a company, social media is an indispensable source of data. Companies actively scan platforms such as Facebook, YouTube, Instagram, Twitter and Pinterest to decipher preferences, choices and perceptions towards brands, companies, political parties and so on. You name it, they want to find out what you think about it. And we give this data away freely and willingly. On Twitter, there are an estimated ten billion tweets per day. Facebook's one billion users generate four new petabytes (about four million gigabytes) of data a day, and four million 'likes' a minute. Just typing in the brandname 'Nestlé' into Facebook will trigger a response back at Nestlé's headquarters.

Even more invasive has been the use of mouse tracking (the ability of companies to remotely record every cursor movement you make on your computer) to gather data on visitors to websites and the development of predictive 'interruptions' to stop viewers leaving a website. Sometimes, you are being tracked even without your knowledge or complicity. A 2008 project called Optic Nerve, a mass surveillance program run by the British signals intelligence agency organisation GCHQ (helped by the US National Security Agency), surreptitiously collected private webcam still images from users while they were using a Yahoo! webcam application. Unsurprisingly, the main discovery was that people were more often than not exchanging pictures of their body parts – an activity classified by the Optic Nerve program as 'undesirable nudity'. It is not known whether Optic Nerve has ceased operations, but since 2014 Yahoo! has offered the option of encrypting all communications within their servers. But the lesson is that you really are never alone when you are using your computer – or any other device connected to the Internet, for that matter.

Transactional data, generated from the likes of eBay, Amazon, your credit card provider and any online retailer you visit, is gathered by the bucket-load and stored in vast files. It is primarily used to predict your consumer behaviour. Target, the large US retailer, can now accurately predict when one of its customers will have a baby just by knowing their personal data and mapping it to their expected life cycle. In China, a 2013 movie called *Tiny Times* was apparently a surprise box office smash – but the producers knew they would have a hit. Social media data was used to choose the stars, director and marketing campaign to appeal as much as possible to the younger generation. The film took around $11.9 million at the box office on its first day alone.

With Big Data getting bigger by the second, storage is rapidly becoming a problem. It's only natural to ask whether there is a limit to the amount of data we can generate and store?

The trouble with tribbles

In the pre-digital age, most data was stored on videotapes, long-playing vinyl records, audio cassettes – generally things you could hold in your hand, lose, pour drinks over, or set on fire. The flammability of human knowledge is exemplified by the estimate that 33 per cent of all human knowledge was stored on paper until as late as 1986. In stark contrast to the 1980s, a staggering 94 per cent of our data was being stored in a digital format by 2007. Between 1986 and 2007 the amount of data stored digitally was growing at an annual rate of 23 per cent.

It hasn't slowed down since. For this reason alone, any estimate of today's data storage will be out of date immediately – an estimated 2.5 exabytes (2.5 billion gigabytes) of data is generated every single day worldwide. By 2020, data scientists predict there will be 40 zettabytes (400 billion gigabytes) of data in existence; only a few years ago the entire World Wide Web contained a mere 500 exabytes (500 billion gigabytes).

This leads to the problem of storage. Big Data's deep-seated storage problems are illustrated by an episode of *Star Trek* called 'The Trouble with Tribbles'. The Starship Enterprise has taken on board small, furry, puff-shaped animals called Tribbles that purr contentedly – a sound the crew finds soothing. All goes well until the Tribble population starts to double in size. That's not the problem, though – it's the rate they are doing it which is troublesome. It's happening every few hours. Soon, Captain Kirk is left scratching his head while the Tribbles swill around his ankles. Fortunately, and before an unnamed red-shirted security officer chokes on one, the Tribbles manage to poison themselves on adulterated grain due to their gluttony.

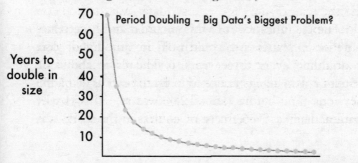

What 'The Trouble with Tribbles' illustrates is a phenomenon called Period Doubling, which can be restated as a question: 'At a given growth rate, how long will something take to double in size?' This is a very real problem and is, by and large, a problem of pure mathematics – whether you apply it to Tribbles, data, the human population, the interest earned on the money in your bank account, or 'things' in general, the answers are always the same. For a growth rate of 1 per cent per year it will take something about seventy years to double in size. When the growth rate increases to 5 per cent, then it takes about fourteen years to double in size. When you get into double-digit growth rates, it takes only seven years to double in size. However, (and here comes the problem for Big Data) when you get to the kind of growth rate being experienced by Big Data – 23 per cent per annum – it takes only three years to double in size.

The significance of this mathematical certainty can't be emphasised too much. If the amount of data is doubling every three years, it equates to the same about of data being generated in three years as all of the previous time before (about 2,000 years). This is why statements like '90 per cent of all data in the world was produced within the last two years' are unarguably, but alarmingly, true.

To cope with the 2.5 exabytes of data we generate each and every day (2.5 billion gigabytes), Amazon, Google, Facebook and Microsoft now have, at the time of writing, over 4 million servers (computer hard disks) between them, throbbing away day and night absorbing data – and the numbers only grow by the day. In 2012, Facebook was already using as much electricity at its data centre in Oregon's Crook County as the rest of the 26,000 residents of the area. Since then, its energy usage may have doubled. Some people believe that the growth of data will one day see us having to pay to upload photographs or interact with the Internet as a form of environmental tax. In many respects, one of the biggest challenges for Big Data in the twenty-first century is not how we expand storage capacity, but how it becomes 'Small Data' through more advanced compression techniques and even rationing of our access to the Internet.

Big Data vs Big

The first human being captured in a photograph was a man on the Boulevard du Temple in Paris having his shoes polished. Louis Daguerre snapped the unsuspecting trailblazer in 1838. Ever since, we have been on an ever-escalating race to record our personal narratives and broadcast them to the world. Combined with the invention of the Internet and the ability to track and record our every journey, website visit, mouse movement, expression of personal preference and even opinion, it is small wonder that the digital data universe is growing at a rate of over 50 per cent a year. It's also small wonder it is the most common source of anxiety amongst Internet users. In a 2015 survey, the main concern of users about the data being collected on them by companies was: 'If/Where they sell my data' (78 per cent of 2,062 people).

You can now watch the Internet grow in real time at places like www.webpagefx.com/internet-real-time. While globally the number of users has passed the three billion mark, India has surpassed the US as the world's second largest Internet market. In first place is China, which boasts leading-edge companies in online commerce, messaging, travel and financial services.

With global expansion has come an explosion of data generated by a multitude of devices. Fortunately, while the amount of data in the digital universe expanded ten-fold between 2010 and 2015, the cost of storing it declined from about $0.20 to just $0.02 over the same period. We now live in the age of unfettered data production, storage and even hoarding.

Big Data vs Analytics

Does 'The Ballad of Bilbo Baggins' by Leonard Nimoy define our future?

For a statistician, John Graunt had a suitably uncertain beginning. He was born in London in 1620, as one of seven, or maybe eight, children. Graunt started out as a haberdasher but fled the trade following the death of his father to 'become influential in the City'.

However, this wasn't enough for Graunt. With his friend William Petty, an economist, anatomist and statistician, he laid the foundations of 'demography', enabling the calculation of survival rates at each stage of human life based on probability. Seventeenth-century London was a place of unusual peril – each day could be your last, to a degree we couldn't imagine today and for causes that seem minor. Each year, several citizens died of 'lethargy', while in 1660 nine people perished of being 'frighted'. It's enough to make you squirm in your seat to imagine the fatal writhings of the one person Graunt recorded as dying of 'itch' in 1648. Each one of these deaths was carefully recorded from the 'bills of mortality' (weekly statistics of deaths) and analyzed.

Graunt's book, *Natural and Political Observations Made upon the Bills of Mortality,* ran to five editions by 1676.

Although they weren't successful in their main aim of creating an early warning system for the onset of bubonic plague, through their careful collection and analysis of the causes and timing of the passing of Londoners, Graunt and Petty created the 'life tables' (which are, in fact, 'death tables') used by insurance companies all over the world to this day. Graunt was elected a fellow of the Royal Society in 1662 for his troubles.

Graunt and Petty worked in a way that today's data scientists would recognise. They established the idea that collecting data and applying advanced mathematics could reveal hitherto unforeseen truths upon which we could act, either for the greater good or for personal gain. In the process, he also set in motion an almost insatiable demand for data and an even more voracious desire for more complex ways of analysis and visualization of the results. It has been a titanic struggle requiring enormous leaps forwards in mankind's data and analytical capabilities. There have been long periods of frustration when not very much happened, but each time a jump came it was large and world-changing. Every time it seemed we were about to be overwhelmed by the problems, we have discovered new ways to gather data, store it and interrogate it. The expansion of data has revealed layers of complexity hitherto hidden from us, which in turn required more data to be gathered and analyzed.

For this reason alone, to understand Big Data we need to know how it is stored and analyzed. It's the story of how the number crunching of the 1880 US

Census was reduced from eight years to 'just' six in 1890 and how music-streaming services will somehow recommend your next listening pleasure as 'The Ballad of Bilbo Baggins' by Leonard Nimoy. It's really the story of how databases came to exist and how artificial intelligence got into our lives – for ever.

It's not what you do it's the way you that you do it...

Herman Hollerith was precocious by any measure. He had his degree in mining from the City College of New York by the age of nineteen. At twenty-two he was teaching at the prestigious Massachusetts Institute of Technology (MIT). Considering what he would do next, taking a degree in drilling holes in things was an ideal start.

Working at the Census Bureau, Hollerith's brain wave, encouraged by his boss, John Shaw Billings, was the idea that data could be recorded by the presence or absence of a hole at a specific location on a piece of card. Think of the playing of a note on a mechanical piano roll, but instead of playing a B-flat, the system would register 'married' or 'not married'.

Hollerith's insight was that data stored at a specific location on one of these 'punch cards', arranged in rows and columns, could be counted or sorted by machines rather than by rows and rows of time-consuming human beings with pens and ledgers. If a series of his machines were arranged in the right order, passing the cards

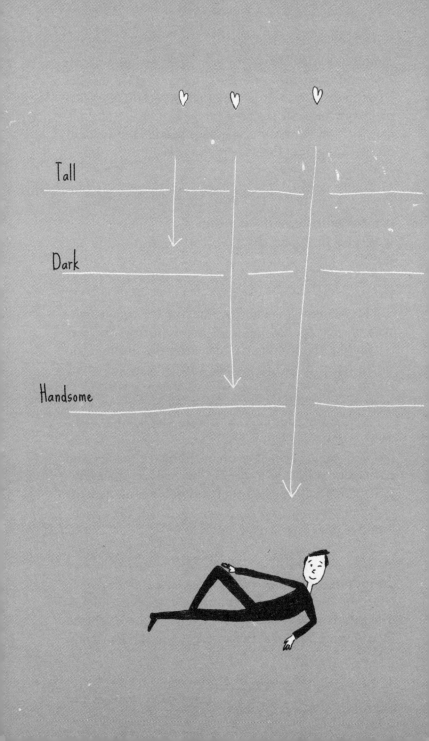

Tall

Dark

Handsome

from one to the other and effectively asking different 'questions', mountains of data could be extracted at rates of 100–2,000 cards per minute. His clicking electrical gadgets were known as tabulation machines.

So powerful was the technique that it spread rapidly around the world. Punch cards were adapted by census bureaux, insurance companies and large corporations. Hollerith went on to found the Tabulating Machine Company, which eventually merged with two other companies to become International Business Machines in 1924 – or IBM, as we know it today. Even though it might appear primitive today, the punch card system dominated statistical data manipulation for the next eighty years, right up until the 1970s.

Punch cards mutated into long pieces of paper tape with holes and eventually great spools of magnetic tape (much like tape recorders) containing digital zeros and ones. However, it was still frustratingly time-consuming and cumbersome to get at the data. You couldn't dip into the middle of a spool of tape without winding through everything that preceded it. Retrieving the information was entirely dependent upon how it was physically organised. There had to be a better way and an impetus to do it. That impetus was eventually the US space programme – and the better way was the computer hard drive.

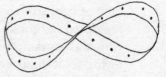

To infinity and beyond!

The Saturn V and Apollo programmes of the 1960s and 70s needed a way to track the vast amounts of materials and components it used. IBM had the answer with hard drives the size of a beer keg, weighing around 900kg (2,000lb) each, which were originally designed for real-time accounting for business. Each drive held between 3.75 and 5MB of data. Today, an average smartphone can hold 3,200 times more information and weighs around 100g (¼lb).

Rather than inflexible tape systems, it had long been realised that a better way to store the data involved in massively complex projects was to use computers and filing systems organised in a giant tree structure. A skilled operator could navigate their way through, starting at the top and gradually finding their way through the branches to locate what was needed. Two rival systems appeared: the hierarchical method and the network method. Essentially, they allowed the user to drill down through the data and even across from different categories of files, depending upon how many connections could be made between the categories. This created yet another problem: the user had to know how the filing system was organised and had to be able to write computer programs in specialist languages to navigate through the system. If there was no road map or driving skills, there was no getting from A to B.

The big breakthrough came when British mathematician Edgar F. Codd threw out the idea that only specialists who knew how the information was organised and could write code could use a system. Codd, who was working for IBM, wrote a paper at the time saying so. It didn't go down well with his peers, because it cut across established IBM businesses. From Codd's rigorous mathematical and logical analysis, he proposed that all data should be held in tables of columns and rows linked, or 'related', to each other by common columns. He called it a relational database, where all the user needed to do was ask the right questions using a logical system called Structured Query Language (SQL) to get at what was wanted.

IBM hated the idea and only grudgingly and slowly adopted Codd's revolutionary thinking. Codd, who was an interesting and slightly eccentric character, became frustrated with IBM and with companies who were jumping on his bandwagon but weren't actually creating relational databases with the purity of his original conception. To his mind, they were merely glossing up what they already had. In response to this, Codd proposed twelve rules ('Codd's Twelve Commandments' as some quipped – and there were actually thirteen rules, as they were numbered from 0 to 12) which defined what a relational database was. Many of them are lost to the casual observer, but one in particular is important because it paves the way for the next revolution:

> **Rule 11:** *Distribution independence:*
> The end-user must not be able to see
> that the data is distributed over various
> locations. Users should always get the
> impression that the data is located at
> one site only.

This is an interesting idea. It creates the environment for Big Data to be viewed not as something to be owned and organised by someone in a little puddle of a hard drive or server; the user can now fish around in an entire data lake, as long as the data is connected by a common field or fields. More importantly, it speeds things up – a lot. And here's why.

Ordinary 'flat file' databases organise information in rows and columns (like a spreadsheet does today). Combining them can make the database long, wide and cumbersome. Codd's vision of the relational database imagined the data split into smaller sub-tables which could be called up quicker and interrogated. Take this example of a flat file database containing information about the users of an imaginary music service showing what they listened to and what they listened to next.

ID	First Name	Surname	Age	Gender	Song	Artist	Genre	Listen to	Next Song	Artist
ID0098	Stewart	Cowley	55	M	Red Shoes	Kate Bush	Indie	15	Wrecking Ball	Miley Cy
ID0098	Stewart	Cowley	55	M	Ships in the Night	Be Bop Deluxe	Prog Rock	10	Even Now	Barry Manilo
ID0098	Stewart	Cowley	55	M	Brighton Rock	Queen	Rock	5	Deeper than Crying	Alison Krauss
ID0099	Bill	Nelson	60	M	Ships in the Night	Be Bop Deluxe	Prog Rock	7	The Ballad of Bilbo Baggins	Leonar Nimoy
ID0099	Bill	Nelson	60	M	Money	Pink Floyd	Rock	8	Across the Universe	The Beat
ID0099	Bill	Nelson	60	M	Shaddup You Face	Joe Dolce	Comedy	15	Everyday Hurts	Sad Ca
ID0102	Ronnie	Woodhouse	57	M	Ships in the Night	Be Bop Deluxe	Prog Rock	12	The Ballad of Bilbo Baggins	Leonar Nimoy
ID0102	Ronnie	Woodhouse	57	M	Ernie	Benny Hill	Comedy	8	Brown Sugar	Rolling Stones
ID0102	Ronnie	Woodhouse	57	M	Bassoon Concerto	Weber	Classical	10	Da Funk	Daft Pu

A relational database splits the information into separate tables, each having a relationship (column) common to each other called a 'key field'. If you wanted to find out what Ronnie Woodhouse (ID0102) listened to most often, the information to call up would be the first two tables rather than calling up all the information as you would have to in a flat file database. The answer would be 'Ships in the Night' by Be Bop Deluxe, which he listened to twelve times.

ID	First Name	Surname	Age	Gender
ID0098	Stewart	Cowley	55	M
ID0098	Stewart	Cowley	55	M
ID0098	Stewart	Cowley	55	M
ID0099	Bill	Nelson	60	M
ID0099	Bill	Nelson	60	M
ID0099	Bill	Nelson	60	M
ID0102	Ronnie	Woodhouse	57	M
ID0102	Ronnie	Woodhouse	57	M
ID0102	Ronnie	Woodhouse	57	M

ID	Song	Artist	Genre	Listen to	Next Song	Next Genre
ID0098	Red Shoes	Kate Bush	Indie	15	Wrecking Ball	Pop
ID0098	Ships in the Night	Be Bop Deluxe	Prog Rock	10	Even Now	Pop
ID0098	Brighton Rock	Queen	Rock	5	Deeper than Crying	Country
ID0099	Ships in the Night	Be Bop Deluxe	Prog Rock	7	The Ballad of Bilbo Baggins	Comedy
ID0099	Money	Pink Floyd	Rock	8	Across the Universe	Pop
ID0099	Shaddup You Face	Joe Dolce	Comedy	15	Everyday Hurts	Pop
ID0102	Ships in the Night	Be Bop Deluxe	Prog Rock	12	The Ballad of Bilbo Baggins	Comedy
ID0102	Ernie	Benny Hill	Comedy	8	Brown Sugar	Blues
ID0102	Bassoon Concerto	Weber	Classical	10	Da Funk	Techno

Next Genre	Next Song	Artist
Pop	Wrecking Ball	Miley Cyrus
Pop	Even Now	Barry Manilow
Country	Deeper than Crying	Alison Krauss
Comedy	The Ballad of Bilbo Baggins	Leonard Nimoy
Pop	Across the Universe	The Beatles
Pop	Everyday Hurts	Sad Café
Comedy	The Ballad of Bilbo Baggins	Leonard Nimoy
Blues	Brown Sugar	Rolling Stones
Techno	Da Funk	Daft Punk

The tables can also be joined to ask more complex questions: 'What do men aged between fifty-five and sixty listen to after they have listened to 'Ships in the Night' by Be Bop Deluxe?' The answer, more often than not (two out of three occasions), is 'The Ballad of Bilbo Baggins' by Leonard Nimoy. That imaginary music service might then say to a customer who hasn't played the Nimoy track something along the lines of, 'Now you have listened to 'Ships in the Night' by Be Bop Deluxe, would you like to listen to 'The Ballad of Bilbo Baggins' by Leonard Nimoy?' – to the customer's surprise, but not necessarily delight.

It may seem a simple change at first but the move from a table format to one with lots of tables that were related to each other had a profound effect on how databases worked. Fragmentation of data sources added a layer of flexibility which would speed up databases many times over. Another step on the journey of speed and fragmentation had been achieved.

The object of your desire...

The 1980s ushered in excessive consumerism, big hair and enormous shoulder pads. It was also the age of desktop computing, empowering users with spreadsheet programs such as Lotus 1-2-3 and database management software such as the pleasingly accessible dBASE. Everything pointed to a sea change away from the drudgery of what had gone before – users could focus on what they wanted to achieve.

'A lot of the dirty work had already been done,' said the creator of dBASE, C. Wayne Ratliff, when interviewed about it a few years ago. 'So the user can concentrate on what he is doing, rather than having to mess with the dirty details of opening, reading and closing files, and managing space allocation.'

dBASE became one of the top-selling software titles in the 1980s and early 90s, which coincided with the emergence of the Internet and the data abundance it brought with it.

Users were realizing that relational databases are fine for some uses, but when the data starts to get really big, you have to find different ways of accessing it. The 1990s saw a new leap forwards in how data was handled. Programmers and designers began to treat data as 'objects' rather than interrelated tables – the so-called 'object-orientated database'.

The simplest way to think of an object-orientated database is to imagine we are going to store all of the work of a group of authors. In the object-orientated world, each author has their own room where all their books are stored. You can even do calculations on the data in the room and post them on the door – how many pages in each book, how many times does the author use the word 'Bismark', and so on. Retrieving the data becomes much faster, because you go straight to the author rather than having to sort through everybody else first.

Using 'objects' was a revolution in itself, but it was still a time-consuming process, as all the searching was going on in sequence, one after another. What if you could give out parts of the search so that multiple computers could work on the problem in parallel – at the same time?

This is where Hadoop came from in 2005. It was borne out of a simple problem: how do you index the entire World Wide Web so you can look things up in a fraction of a second – as Google was trying to do. Yahoo!'s Doug Cutting had the answer, which he named after his son's toy elephant. Hadoop abandoned the idea that a single piece of data should be stored in one place (such as your PC).

To imagine how Hadoop works, it's important to understand how it distributes data to other computers to be analyzed. Imagine just the box of a PC – it contains a processor, some memory and a memory disk. In the Hadoop world, this is called a 'node'. Now stack lots of these PCs on top of each other. This is called a 'rack'. If you get lots of racks in a room, it's called a 'cluster'. Whenever you see rows and rows of computers blinking away at each other, that's a cluster.

Hadoop takes data and copies it in equal-size blocks to a number of nodes simultaneously using its Distributed File System called MapReduce. The programmer uses the Java programming language to write lines of code to tell each node what part of the analysis to work on (the map), then return the results in the form of Key Value Pairs. The Key is the name of the thing you are counting and the Value is how many times it occurred. The Key Value Pairs are sorted and then totalled (the reduce) to get the final result. For instance, if you were searching a years' worth of tweets on Twitter for keywords, you could send four fragments of the data to four nodes and then use MapReduce to search for specific words in each of the individual nodes. This way, everything is running in parallel rather than sequentially, which means it is much faster. Also, you will get a result back – even if one of the nodes fails during the analysis, the others will return their findings. If something goes wrong during the calculation with a standard database, the whole process falls over. Hadoop offers a more reliable way of performing large complex calculations quickly.

The history of databases is ultimately the history of our desire for flexibility and speed. From the careful manual collection of death statistics by John Graunt and William Petty, through Hollerith's card-shuffling system, to today's Hadoop, we have now arrived at the point where data sources can be disparate, separate and very large, but can be investigated very quickly. All you need is something to apply it to.

Neural networks

Neural networks attempt to mimic how human beings think and react by using data combined with advanced analytics. It works something like this: imagine I poke you with a pen. You shout 'Owww!' The input is the pain, the output is the 'Owww!' In between the prod and the exclamation are lots of hidden processes occuring inside your body. Some are more important than others, but they are all triggered by the initial prod. If we wanted to write a computer program to predict that you would shout 'Owww!' when I prodded you, we could if we had lots of examples of author/reader interactions and worked out which of the connections in your brain were the most important. We would give the most weight to those connections that produced a result closest to 'Owww!' If we run the program again and the result was a 'Wooo!' we know it's slightly wrong. By comparing the results we can tweak the weights of the most important process and run the program again and again until the error in the output compared to what we want is minimised. This iterative process is called 'backpropagation' and is how neural networks function. Feeding back via backpropagation might require inputting thousands, if not millions, of examples of what you are attempting to produce. Looking at a visual representation of a neural network can give you insights into actually what is going on in any process or system.

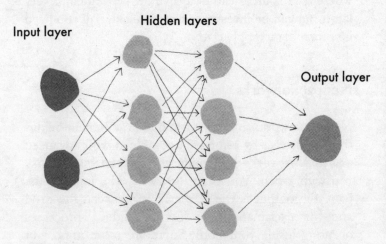

In a neural network, each processing point is called a node and they are arranged in layers. Nodes may go on to trigger another layer, or many layers, of nodes, before the eventual emergence of the result. Google uses neural networks to process language and uses thirty layers of nodes. More importantly, it's how we imagine the human brain works – the connections in your brain called neurons transmit pulses along them and through the chaotic triggering process helps you recognise patterns and make sense of the world. Over time and lots of iterations we become good at recognizing the pattern. It's why musicians practise scales obsessively – each repetition strengthens their neural network, which they can call upon when sight-reading music or in performance. For this reason neural networks are in the category of artificial intelligence, which has been the stuff of science fiction and speculation for hundreds of years. Think HAL 9000 in *2001: A Space Odyssey*, Rachael in *Blade Runner* or Samantha in *Her*.

Neural networks had a bright start, but soon they ran into problems and ground to a halt in the 1980s because the computational power required wasn't available at the time. Since the 1990s interest has grown again, as has the complexity. Modern neural network projects typically work with a few thousand to a few million neural units and millions of connections, which is still several orders of magnitude less complex than the human brain and closer to the computing power of a worm. However, the applications of neural networks have been diverse – from finding out if the addition of a football player will improve team performance before you hire them, to working out the next twitch in the financial markets, or creating expert systems to help detect certain types of cancer. Neural networks are behind the recognition software for voice, face, handwriting and fingerprints – anywhere a system asks for an example in your life of something real about you, there is a neural network humming away in the background. It has diverse commercial uses too. A group at the Punjab Technical University in India have used a neural network to predict with 93.3 per cent accuracy whether a Bollywood film would be a 'flop', 'hit' or 'super hit', based on the actors, directors and music used even before a single frame has been shot.

Neural Network

Machine Learning

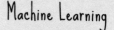

Neural Network	Machine Learning
Aims to emulate human intelligence	Aims to be statistically reliable
Interested in the process and result	Interested only in the result
Does not come close to matching the complexity of the human brain	Still needs to improve on how it deals with voice related commands
Learns in both a supervised and unsupervised way	Learns like neural systems do but also through reinforced learning
Integral to recognition software, predicting human reactions	Integral to detecting fraud, medical diagnosis, online advertising
Behind any system asking for something real about you	Behind any system asking for information about you

One ring to rule them all...

Machine learning came out of the same route cause as neural networks: the quest for artificial intelligence. There is a difference between neural networks and machine learning, however. Neural networks try to understand the world through connections such as the ones we might have in the brain. We tinker with the modelling process (called supervised learning) or let it go (unsupervised learning), but either way, data scientists remain intensely interested in what is happening to produce results. The 'how' is just as interesting as the answer itself.

Machine learning has no such interest. It doesn't understand anything, nor does it want to, because it does not seek to emulate human intelligence. Machine learning looks at data and then makes predictions about the future. It judges its success by how accurately it predicts the next data point in a series by comparing the prediction with what actually happens. It can learn as it goes along in a supervised, unsupervised way (just like neural networks) or by reinforcement learning (a system of rewards and punishments for completing or failing to complete a task). By honing down the difference between prediction and what actually happens by feeding it with more and more data – in the jargon, 'minimizing the loss function'– a machine learning model builds a picture of the world that is statistically reliable.

An area that is often confused with machine learning is data mining. It is closely related to machine learning and is a branch of data analysis. Both use vast data sets in databases analyzed using statistics and the mathematics of probability, but where machine learning looks to reproduce known relationships and predict the future, data mining digs deep into the data to find previously hidden trends or anomalies. This is sometimes called Knowledge Discovery in Databases (KDD). KDD is interested in the odd and the new. Data mining has been used to spot changes in business trends, to classify music into correct genres and even to predict whether an air passenger could be a likely terrorist before they board a plane.

Like data mining, the applications of machine learning are many and diverse. Anywhere there is a lot of data stored in a database and the opportunity to run a program repeatedly, combined with desire to predict the future, you will find machine learning. Gaming, fraud detection, medical diagnosis, online advertising, stock market analysis – anywhere pattern recognition is possible and desirable, you will find machine learning. A pet store is currently developing a machine learning system to recognise pets and match them to their owners and their previous transactions when they enter a store, so sales assistants can be alerted and greet them accordingly. 'Good morning, how is Fido today? It's been six weeks since he last had his worming tablets…'

The biggest growth area for machine learning is in creating a new way of interacting with computer systems. We increasingly talk to them and ask them questions in natural language, such as, 'What was the patient's blood pressure the last time she visited?' or 'I am self-employed, is my van insurance tax deductible?' Since 2008, queries associated with voice-related commands have increased more than thirty-fivefold since the launch of the iPhone and Google Voice Search. Most of these technologies are mature, but they are still in their infancy, despite penetrating every aspect of our lives. The combination of data, databases and analytics will change virtually every aspect of society in the remainder of the twenty-first century.

Google Trends

If you want to play with the power of global database technology and combine it with analytics for yourself, you can. Google Trends (available at trends.google. co.uk) allows anyone to type in keywords to see how often a particular search term crops up over time. This can be performed in real time or using long-term historical data.

The numbers require some interpreting because they index to 100 – which is the time the peak interest occurred. For instance, if you want to watch the relative rise and fall of TV shows *The X Factor* versus *Strictly Come Dancing* in the past five years it's clear *The X Factor* 'peaked' in 2012 and since then it has been a downward spiral of seasonal popularity. Meanwhile, *Strictly Come Dancing* has risen in the ratings and is now pretty much neck and neck with *The X Factor*.

Similarly, if you were one of the many who mourned the passing of singers David Bowie and George Michael in 2016, you could hardly miss the outpouring of grief for these artists. To give a sense of proportion, though, it was nothing compared to the shock and grief displayed on the Internet when Michael Jackson died.

By comparison, David Bowie and George Michael made hardly a ripple in the fabric of the data universe.

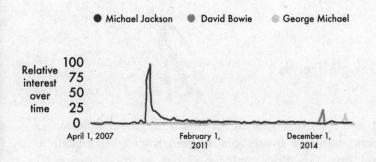

It's not only trends in entertainment that are searchable. A wide range of social, economic and financial market data is also available. Google data scientist turned *New York Times* op-ed writer Seth Stephens-Davidowitz used Google Trends to show how President Obama lost about 4 per cent of the vote because of racial prejudice in the 2008 election. Others have used Google Trends to correlate the names of companies and the corresponding volume of stock transactions in the coming weeks. There has even been evidence produced that if you are wealthy and have access to the Internet, you are more likely to be seeking information about the future than looking back.

Big Data vs Analytics

In 1872, author Samuel Butler imagined the evolution of machine consciousness in his anonymously published novel *Erewhon*, even though the term 'artificial intelligence' (also known as AI) was first used in 1956 by computer scientist John McCarthy. The well-known British scientist Alan Turing had the foresight to predict that a computer would be able to imitate human thought by the year 2000. He was only three years out: in 1997, IBM's Deep Blue beat chess champion Garry Kasparov, much to his despair. In 2016 an IBM algorithm called ROSS became the world's first AI attorney, landing a job at New York firm Baker & Hostetler in its bankruptcy practice. We've clearly come a long way since John Graunt scoured the weekly 'bills of mortality' for glimpses of the future.

Needless to say, not everyone is happy about some of these developments. The use of data and advanced analytics can be seen as a benefit while simultaneously attracting concern. The benign results are better and more timely diagnostics of medical conditions, or the United Nations' HunchWorks program, which combines diverse data sets from social media, YouTube and Internet searches to detect crises in their very earliest stages. Companies can deliver better customer experiences and adjust products and even pricing using real-time feedback. Japan has invested heavily in robotics to tackle the problem of its aging and declining population. In the three years up to 2016, Japan doubled the number of intelligent robots in its industrial base, sometimes slashing labour costs by 80 per cent. On the face of it, a glorious evidence-based world lies ahead of us.

However, like all systems, it is impossible to avoid the idea that if you put rubbish in, you will get rubbish out. Machine learning and data mining in particular can be worrying to some. If the data going in is institutionally biased for or against a particular group in society, then it will produce conclusions reflecting the bias. It can also throw up connections that are tenuous at best, or may not even be real. The Computer-Assisted Passenger Prescreening System II (CAPPS II) and its successor, Secure Flight, which used information stored in government and commercial databases to assign a colour-coded risk level to passengers, generated civil liberties and privacy concerns because of the false positives that were thrown up. In one such example in 2004, CAPPS II refused boarding to Senator Ted Kennedy.

Creating new AI systems brings with it an ethical and moral responsibility for data collection when implementing supervised and unsupervised artificial intelligence or machine learning systems. For instance, there are human resource systems designed to predict whether an applicant for a sales job will, even before they are employed, hit their targets just from the answers on a questionnaire. If the data used to create the accept/

reject model in the system has 'bad experiences' with one type of demographic, it may discriminate unfairly over able candidates, denying them employment.

Then there is the nagging worry that, one day, machines will take over the world. So far, machine learning and AI has only been applied to very narrow problems: playing a game, solving a problem, and so on, rather than allowing it to roam freely, which experts in the field of ethics currently find comforting. Or maybe it's just that we haven't asked one to 'Go and take over the world'. Yet.

Big Data vs Privacy

Is your toaster talking to
your fridge about you
behind your back?

'The first rendezvous attempt will be at 10 a.m. on Monday. We will meet in the hallway outside the restaurant in the Mira Hotel. I will be working on a Rubik's cube so you can identify me…' So it began. In June 2013, Edward Snowden, a twenty-nine-year-old United States National Security Agency (NSA) infrastructure analyst with privileged access rights to classified information, met with Glen Greenwald, a UK-based journalist working for *The Guardian* newspaper, in a hotel in Hong Kong. It was a low-tech, twentieth-century process, somewhat out of kilter with their high-tech twenty-first century subject: the mass surveillance of entire nations using computers, databases and advanced analytics.

Snowden revealed the existence of mass surveillance by the NSA and the UK's Government Communications Headquarters (GCHQ) with the complicity of other governments. Most shocking was the involvement of corporations such as Apple, Facebook, YouTube, Skype, Yahoo! and Microsoft – companies consumers trusted with their data. Some of the cooperation dated back to 2007. Worst of all, Snowden confirmed the fears of the Internet fringe at a stroke and made true Joseph Heller's *Catch-22* maxim, 'Just because you're paranoid, doesn't mean they aren't after you'.

Edward Snowden opened a Pandora's box of information. He may have revealed tens or hundreds of thousands of documents (depending on who was estimating the total) filled with acronyms and code names such as PRISM, XKEYSCORE, UDAQ, Tempora, FISA, and MINARET. The documents revealed a systematic invasion of privacy through email, video, photographs, stored data, file transfers, video conferencing, logins and social network surveillance, all done in the name of national security in a post-9/11 world.

Some hailed Snowden a hero, while President Obama described his methods as 'damaging'. Is our privacy a small price to pay for our security? If you want to stop the snoopers, what can you do to stop them in the Big Data era? To tackle these questions, you have to understand what is going on now and how we got here.

Oh (Big) Brother, where art thou?

Prior to 9/11, the efforts of national governments to gather information on their citizens was, by comparison with today, frankly amateurish and of a scale that bordered on the dilettante. In the US, Project MINARET, operated by the National Security Agency, intercepted the electronic communications of pre-designated US citizens and passed the information on to the FBI, CIA and the Secret Service. No legal warrants were needed and there was no oversight of the 3,900 citizens targeted. In 1972, legislation stopped MINARET in its tracks as the United States Supreme Court upheld the Fourth Amendment of the Constitution of the United States which prohibits unreasonable searches and seizures without a warrant, even if someone was trying to overthrow the government.

Among the results that came from the aftermath of MINARET was the creation of the Foreign Intelligence Surveillance Act (FISA), intended to limit the powers of the NSA and which required a process, including warrants and judicial review, before surveillance could begin. The events of 11 September 2001 kick-started the warrantless mass gathering of data, when the administration of President George W. Bush assumed unitary authority – the theory that American presidents can do pretty much anything they like – and thus began the 'Stellar Wind' project under the President's Surveillance Program (PSP). Stellar Wind wasn't entirely successful; the cases referred to it by the FBI became known as 'pizza cases' by agents, because many seemingly suspicious connections turned out to be food takeaway orders. What they needed was someone with the right skills to make the right connections.

William Binney was that guy with the right skills but he was, arguably, in the wrong place. Binney was a career employee at the NSA with an astonishing aptitude for mathematics and code breaking. After a brilliant career during the Cold War, in the 1990s he turned his attentions to the Internet. Binney co-founded a fledgling unit automating signals analysis that drew together data from lots of sources at the same time to construct profiles. When American Airline's Flight 11 struck the northern side of the North Tower of the World Trade Centre at 8.46 a.m. on 11 September 2001 not only was there a reason to increase surveillance, but Binney was primed and ready to go. His own recollection was it took less than a week for him to be contacted by the NSA to supercharge his programme.

Within the NSA, the programme ran under the code name PRISM. To help it along its way, the US Intelligence Budget rose from \$27bn in 1998 to \$72bn in 2017. Even allowing for inflation, that's a funding increase of 180 per cent. Some \$11bn of this may now go to the NSA alone.

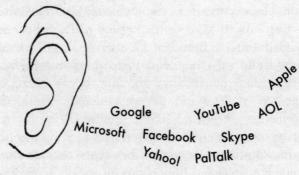

As PRISM grew in scale and scope, Binney became increasingly disturbed by what he was witnessing. Eventually, he turned whistleblower over the legality of the NSA programme. In the process, he revealed that the Bush administration had authorised the collection of domestic communications running into hundreds of millions of interceptions per day. PRISM collected (and by all accounts, still collects) Internet communications from at least nine major US Internet companies, including Apple, Alphabet Inc (Google), Facebook, Amazon and Microsoft (including Skype). XKeyscore is the front-end program designed to filter keywords and to identify targets. Just by searching for a trigger word, it is possible you could be labelled an 'extremist'.

The daddy of all surveillance efforts is probably from a British GCHQ programme called Tempora (the name possibly comes from the Latin phrase '*tempora mutantur, nos et mutamur in illis*', meaning 'Times change, and we change with them'). Since 2011, Tempora has been intercepting fibre-optic cables – which make up the backbone of Internet and telecommunications infrastructure – to access users' data without them ever knowing. Like PRISM, Tempora has a reach that is both domestic and international. It operates with the full knowledge of the companies the hardware belongs to.

According to *The Guardian* newspaper, Tempora is something of a joint project by GCHQ and the NSA and they both have several hundred analysts working on it. The interceptors are placed on some 200 fibre-optic cables where they come ashore along the coast of Cornwall in the south of England. This gives GCHQ access to 10 gigabits of data per second, or 0.864 petabytes a day. A technique called Massive Volume Reduction (MVR) is used to discard high volume, low-value traffic, thus reducing the volume of data to be processed by 30 per cent. Specific searches, trigger words, email addresses, or targeted persons, and telephone numbers bring the volume down further. GCHQ and the NSA have identified 40,000 and 30,000 triggers respectively. As things stand, GCHQ generates more metadata (the trail of data crumbs that makes you identifiable) than the NSA. The data is preserved for three days, while metadata is kept for thirty days. Regardless, it is a massive undertaking.

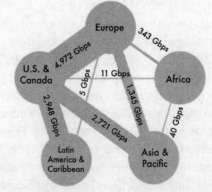

The reason programs such as PRISM and Tempora are so successful is both technical and due to our interconnectedness. The technicalities of intercepting, storing and interrogating data on this scale are intertwined with the history of the database, the increasing sophistication of querying techniques, and the development of vast storage server facilities covering the size of many football pitches. Meanwhile, the process of keeping tabs is helped by the fact that any communication between two people may pass across a number of jurisdictions before the connection is made. As this slide by the NSA on the PRISM program notes:

> *A target's phone call, email or chat will take the cheapest path, not the physically most direct path.*

Your target's communications could easily be flowing into and through the U.S.

This routing process is important, as we are now about to discover.

'Meh...'

You have to dig pretty deep into any website or app to find its privacy policy. You can usually find it tucked away at the very bottom of the home pages behind links marked 'Legal', in very small print. If you do find it, it seems to be hoped that you will give up after a few paragraphs, exclaim 'Meh...' and move on. This is doubly true when signing up to a new service – pressing the 'accept' button when visiting a site or app for the first time circumvents so much suffering for so many people. However, it may just be worth taking a look at what you are signing up to. It would certainly open your eyes.

There are two levels to how websites obtain and use data. First, you have to accept the privacy policy as presented. Second, you have to allow the website to place 'cookies' on your computer. Cookies are small files placed in your browser or on your hard drive by the website. They monitor what you do and feed it back to the website provider and their friends.

They sound cute, but cookies are actually a devil in disguise – and there can be a lot of them sitting on your computer. For one site I use regularly, there are forty-one of these little robots listed in the cookies policy that measure my every cursor movement, page visit and each time I hover the mouse over a certain page. All I'm doing is looking for cinema tickets.

Cookies thrum away in the background; I found over 2,000 of them on my computer when I looked in detail recently. If you open the Privacy Preferences of your browser, you will find an area where you can block cookies or clean them out. A warning, though: some sites simply won't allow you to block cookies. They won't work without them in many cases. You don't have a choice about being monitored.

Any company with a website has to have a privacy policy by law. Fortunately, there are websites that will automatically generate a privacy policy for you (which, ironically, force you to accept their cookies to get at what you need). Even so, there is a remarkable level of uniformity in privacy policies. Opposite is a survey of privacy policies from my life – each one meticulously read. You may like to do the same.

Some interesting things jump out. These agreements explicitly state that cookies will enter your device, have a look at the hardware you are using and pick up the unique identifier for your machine. They will also identify your device model, operating system, its version and much more, all without your knowledge. Cookies will routinely store your latitude and longitude through the GPS system. A number of the sites and apps require that you have your GPS system turned on at all times. This will all be sent back to the company and added to their Big Data store.

Even though many privacy policies explicitly state 'The Internet is not a secure place', they have a remarkable appetite for sending your data over it. Cookies will pick up things that identify you – your email address, login,

Company	Transport data to the US	Share with Third Parties	Forced to comply with government requests	Cookies required to work	Device investigation	Identification data	Data as financial asset	GPS Data
Amazon	✔	✔	✔	✔	✔	✔		
London Live Bus Countdown	✔	✔			✔	✔		✔
Cineworld	✔	✔	✔	✔		✔	✔	✔
Facebook	✔	✔	✔	✔	✔	✔	✔	✔
Jaguar Incontrol	✔	✔	✔	✔	✔	✔	✔	✔
LinkedIn	✔	✔	✔	✔	✔	✔	✔	✔
Great Run Training	✔	✔	✔	✔	✔	✔		
Samsung	✔	✔	✔	✔	✔	✔	✔	✔
Spotify	✔	✔	✔	✔	✔	✔	✔	✔
Google	✔	✔	✔	✔	✔	✔	✔	✔

phone number, and so on. The intention is to customise your experience of the site, but what is also happening is a picture is being built of you to share with third parties so advertisements can be targeted at you. Still, there is no guarantee that third parties will behave themselves – as this common clause points out:

> *Our websites may link to a number of external websites, for example to social networks such as Facebook (so you can login via Facebook using Facebook Connect) and*

> *to our national partners who offer discounts and deals to our Unlimited members. If you click on one of these links you acknowledge that your access of such websites will be subject to that website's own terms and conditions and privacy policy.*

Google goes as far as reading your emails to achieve the same aim, as this clause points out:

> *Our automated systems analyze your content (including emails) to provide you personally relevant product features, such as customised search results, tailored advertising…*

It is understandable that companies will want to scrape as much information when you touch one of their sites – it really shouldn't be a surprise. However, you might expect such actions would be governed by the laws of the country in which you are located. Not so! More often than not, companies include a clause such as this in privacy policies to account for the fact that your data may be spread across many jurisdictions. If you live in Europe don't assume your data will stay in Europe, for instance. This is usually encapsulated in a clause that looks like this:

> *The information that we collect from you may be transferred to, and stored at, a destination outside the European Economic Area ('EEA'). It may also be processed by staff operating outside the EEA who work for us or for one of our suppliers.*

The NSA knew what it was talking about when it stated that, 'A target's phone call, email or chat will take the cheapest path, not the physically most direct path'. Your data is more than likely being transported outside of your home country to companies around the world. Each time it travels down a cable, crosses a junction box or a national border, it is vulnerable to interception. Room 641A in the SBC Communications building at 611 Folsom Street, San Francisco, USA, occupied by telecommunications company AT&T, is a case in point. This room is fed by fibre-optic lines from beam splitters carrying Internet backbone traffic (emails, video, financial transactions) mainly emanating from Asia to the US. Room 641A has access to all Internet traffic that passes through the building and the capability to enable surveillance and analysis of Internet content on a massive scale, including both domestic and overseas traffic. An attempt to shut down Room 641A by the civil liberties group Electronic Frontier Foundation has been quashed by judges.

To address public concerns, the European Union and the United States have now agreed the EU-US Privacy Shield, which both Google and Amazon display prominently in their privacy policies. For the first time, there is a written assurance from the United States that any access by public authorities to personal data will be subject to clear limitations, safeguards and oversight mechanisms. Specifically, Privacy Shield seeks to reassure the public that:

- US authorities affirm absence of indiscriminate or mass surveillance.

- Companies will be able to report approximate number of access requests.
- There will be a redress possibility through the EU-US Privacy Shield Ombudsperson mechanism, independent from the intelligence community, handling and solving complaints from individuals.

Despite this, all privacy policy agreements contain words stating that companies can be forced, by government warrant, to hand over any information they have. Google's transparency report, a laudable initiative to reveal the extent of government requests for information on users, shows that the annual number of user data requests rose from 26,000 to 86,000 between 2009 and 2015. The US, UK, Ireland, Germany and India are the top five requesters. Conversely, if you want to access the information that companies possess about you, it's more than likely you will have to pay to have them reveal what they know.

There are also things in privacy policies that are just plain odd. The motor company Jaguar wants me to read out its privacy policy to any passengers in my car:

> *Please be aware that it is your responsibility to alert all passengers and people you authorise to use your Vehicle or the InControl Services about the privacy practices described in this policy (including the ways in which we may collect and use data from the Vehicle and/or relating to Users of the Vehicle).*

Meanwhile, Samsung has this to say about its voice-activated systems:

> *[We may collect] Voice information such as recordings of your voice that we make (and may store on our servers) when you use voice commands to control a Service. (Note that we work with a third-party service provider that provides speech-to-text conversion services on our behalf. This provider may receive and store certain voice commands.)*

Samsung's Smart TV range of sets have the ability to record what you are saying, even when you aren't giving any commands. In March 2017, Wikileaks – the organisation set up by Julian Assange as a vehicle to publish censored or classified information – disclosed that the NSA and the UK's MI5 had been working on a hack called 'Weeping Angel' that allowed the security services to listen in to conversations through the Samsung F8000 series TV, even though the set appeared to be switched off. It was not disclosed whether other models had been targeted.

His dark materials...

What we see is that the World Wide Web is in fact just the tip of a very large iceberg – the part you would see bobbing along the ocean. Underneath that are places hidden from or not accessible to the general public – government departments, commercial networks in companies – which is called the deep web. Below the deep web is another level, a place only accessible by people with specialist software called the darknet. To escape surveillance, people are increasingly turning to the so-called 'Darknet' – private networks of computers set up to swap files and create content anonymously.

Darknets have been a safe haven for whistle-blowers, content piracy and as a place to trade in things not available in average shops or stores. To add an extra layer of obfuscation, darknets are also places where you can use anonymised virtual currencies such as bitcoin to pay for goods and services. For traders and customers, such platforms are not without risk. Ross Ulbricht, for example, ran a darknet exchange, which he called Silk Road. Part of his original vision for Silk Road was as an economic experiment in free trade. However, his darknet excursion got him into deep trouble; he was sentenced to life without possibility of parole in 2015 for charges including the selling of narcotics, money laundering and 'maintaining an "ongoing criminal enterprise".' However, there is one part of the darknet where anonymity is preserved to this kind of degree: the Tor network.

The popular image of Tor is of shady, masked anarchists preparing for the destruction of the 'Global Illuminati', like a modern-day Guy Fawkes. In fact, The Tor Project was developed by the United States Naval Research Laboratory and in 2012 was 80 per cent sponsored by the US government. Tor works similarly to the way Batman orders his equipment: lots of people make bits of it, but only he knows what the final thing looks like. When a user sends out an encrypted request for a webpage using a Tor browser, it passes through a network of relays randomly selected from several thousand possibilities; each relay layer decrypts a small amount of the original request. The final relay decrypts the innermost layer of encryption and sends the

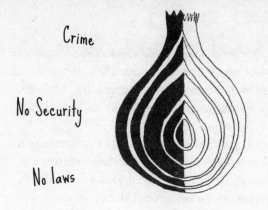

Crime

Freedom

No Security

No Surveillance

No laws

Independence

unencrypted request to the commercial Internet without revealing, or even knowing, the source. As the routing of the communication is partly concealed at every hop in the Tor circuit, this method severely reduces any single point at which an attempt at surveillance can occur. Like the layers of an onion, it is a gradual peeling process and also where Tor gets its name: The Onion Router. Returning the result to the user, the whole process merely goes into reverse.

To access the darknet you will need a Tor browser (based on the popular Firefox framework), which increasingly looks and works like commercial browsers, but at the moment still has the feel of browsers we used in the 1990s. More importantly, the darknet tantalizingly maintains the possibility of a return to the principles set out in writer John Perry Barlow's influential 1996 essay 'A Declaration of Independence of Cyberspace', where the Internet is a free space and an alternative to the established systems of power and money:

> *We are creating a world that all may enter*
> *without privilege or prejudice accorded by race,*
> *economic power, military force, or station of*

*birth. We are creating a world where anyone,
anywhere may express his or her beliefs, no
matter how singular, without fear of being
coerced into silence or conformity.*

If you don't want governments, hackers or corporations
to see what you are doing online for reasons of
anonymity, privacy or security, the darknet is probably
where we are all going to look for that free space.

Five ways to protect yourself and your data

There can be only a small amount of irony attached to
the fact that about five minutes after I typed 'Top ten
tips to protect your data' into Google, an advertisement
with the title 'Protect your Mac' popped up in Facebook.
As we have seen, that's the quid pro quo of the 'free'
Internet: a seemingly endless network of connectivity
between the suppliers of goods and services and social
media platforms. Be careful what you type into your
browser – you will almost certainly be bombarded with
opportunities to spend money.

The problem of protecting your data and privacy is
a very real and present danger. Here are five things – not
a definitive or exhaustive list – that you can do to help
deter unwanted attention.

1. Give yourself a privacy review

It may be tedious, but read the privacy policy of
the services you have signed up for and enable the

information you want the providers to store and share. The rule should be to choose as little data-sharing as is possible. This is best done when you enter a service for the first time. Don't fall victim to the 'meh...' phenomenon. The website and app MyPermissions.org offers a tool to do this.

This applies to advertising and third parties, but should also apply to your social media settings: always choose the highest level of privacy possible to ensure your personal data doesn't end up in the wrong hands and keeps your reputation intact. For instance, on Twitter it is possible to make sure that only approved followers see your tweets, so you can make a separation between public and private sharing. It has never been truer than today that a reputation takes a lifetime to create – and seconds to destroy.

2. Passwords

We are all vulnerable to password theft. In 2012, the social networking platform LinkedIn was hacked and over 100 million logins and passwords were stolen, then offered for sale on the darkweb. The daddy of all thefts in terms of size was the social network MySpace: 360 million usernames and passwords (and second passwords) were downloaded by hackers. When 'qwerty', 'password' and '12345678' are amongst the most highly ranked, but weakest passwords, around, it's actually a wonder that it takes as long as three hours to crack most of them. You certainly shouldn't use the same password for more than one account or service.

If you don't have fingerprint security on your device, here are some tips to create a bulletproof password:

- Use at least twelve characters – the time to crack your passwords increases by hundreds of years of computing time as you increase the number of characters from eight to twelve
- Use a random alphanumeric password generator handled by a password manager that combines upper and lowercase letters with numbers, e.g. 8h$uMP11bCF4
- Use a memorable sentence, e.g. Myfirstpettitch
- Never re-use a password
- Use two-step authentication – this could be in the form of a text with a passcode

Password storage is also a problem. You really shouldn't write them down (I bet you have yours on a sticky note under your mousepad!) except as a hint or clue to jog your memory. Using the 'stickies' program – the electronic version of Post-it notes – on your Mac or storing your passwords in your browser are equally inadvisable.

3. Encrypt your hard drive

Apple's OS X operating system comes with FileVault, a program that encrypts the hard drive of your computer. Microsoft Windows also comes with an encryption program. However, don't assume that encryption is enabled. Go to your system settings to see the state of encryption on your machine. If you do encrypt your hard drive, it won't stop government authorities from demanding your encryption key, but you will deter hackers.

4. Don't give too much away on social media

Your Personally identifiable information (PII) is much more than your credit card numbers, social security number and other details. In the age of Big Data, it's possible to piece together a picture of a person using bits of information taken from multiple sources: email address, browsing history, 'likes' on Facebook and so on. You should be overly cautious about what you share through social media. How would you feel if someone walked up to you in the street and said, 'Hey! Do you go to Harry's Bar every Thursday at 7 p.m.?' Your instinct should be, 'Why do want to know?', 'Who are you?' and 'If you don't leave now, I'm calling the police.' Your default position should be to withhold information, even though social media sites are specifically designed to tell the world who and what you are, when and where you are doing it and who with. Do you have to make it so easy for hackers by having your date of birth registered openly?

5. Switch off Bluetooth when you aren't using it

At a conference called 'Computers, Freedom and Privacy' a group led by Pablos Holman of the Intellectual Ventures Lab set up computers in each room of the venue and logged all the Bluetooth traffic to create a map of the movements of all the delegates. Targets included Kim Cameron, Chief Architect of Access at Microsoft. Bluetooth attacks typically aren't devastating, but can be inconvenient. It's also good practice to switch off your computer or mobile device when you are not using it. At the very least, you should sign out of online services you are not using at the time.

Big Data vs Privacy

There was a time when only a few people lived in the world of Edward Snowden. It was almost a golden age of blissful ignorance for us, but also a golden age for those who wanted to watch us. Now we all live in the same place and there really is no excuse for not knowing about or considering what has been revealed.

I could be forgiven for believing that it's more than likely the act of researching this chapter has set off alarm bells somewhere. I have looked at Wikipedia's pages for 'Edward Snowden', the 'NSA', 'GCHQ', 'PRISM', Tempora, XKeyscore and a whole host of other things including, just to throw them off the trail, the cast members of an old BBC TV drama called *Angels*, as well as many other things my mind wanders off to during research. There may have been a pixel-sized cookie placed in one of the top-secret slides leaked by Edward Snowden and now freely available on Wikipedia. It may have downloaded itself and immediately started broadcasting. My intimate (but shamefully, rather dull) life is now on full display to my GCHQ or NSA operative. It really is going to be dull – nothing I have read has come from anything other than the commercial Internet, the one where if you aren't paying, you are the prey. Still, from all that metadata they must have collected, they could be forgiven for concluding that I had started to look like a threat to national security.

Edward Snowden and William Binney don't disagree with the idea of hunting down and preventing bad guys from doing bad things. They don't even disagree with the methods – they spent most of their careers perfecting them. What they disagree with is the way that it is done: en masse and hidden from the scrutiny and oversight or even against the spirit of existing laws, if not the laws themselves.

Former US President Obama couldn't pardon Edward Snowden because he hasn't presented himself to a court of law and been convicted of anything. However, Obama did offer these words that neatly sum up the balance to be struck between the need for security and the right to privacy in the interconnected, Big Data era:

> *My experience is that our intelligence officials try to do the right thing, but even with good intentions, sometimes they make mistakes. Sometimes they can be overzealous. Our lives are now in a telephone, all our data, all our finances, all our personal information, and so it's proper that we have some constraints on that. But it's not going to be 100 per cent. If it is 100 per cent, then we're not going to be able to protect ourselves and our societies from some people who are trying to hurt us.*

Big Data vs Dating

Is the mathematics of love just 'Love plus One'?

Cave paintings appeared around 40,000 years ago in Europe and Asia. Their exact purpose is not known, but they were found in the areas of caves that we do know weren't usually inhabited. This should be a clue, because amongst the bison and hunting scenes the earliest paintings are hand stencils made by blowing red paint onto an outstretched hand on the wall. This is clearly a dating site.

For the uninitiated this may require some translation. Each handprint is saying something along the lines of:

> *Mature male (16) with excellent lung capacity and large, painted right hand, seeks mature lady (14) for hunting, scavenging and sabre-toothed tiger avoidance. Must have GSOH and own red paint. No time wasters.*

In its early days, the Internet pretty much served exactly the same purpose. Although we wrapped it up in talk of a technology-driven productivity boost in a newly interconnected global economy, the majority of traffic at the time seemed to consist of business people passing lewd jokes between themselves and everyone flirting. The mobile phone could almost specifically have been designed for adultery – everything else can be seen as a byproduct.

It's not surprising that as we increasingly retreat behind our screens, our personal relationships have become more digitised and driven by data. Gone are the days when we used to marry people who we worshipped from afar but lived only a few doors away. We now travel more and leave home for education and experience rather than merely to escape restriction. Developing the 'emerging' adult inside us, the person we become in our twenties, has replaced the rush to marry young. We also allow a series of demands of the idealised partner to form – a potential mate must 'complete me' and 'be my soulmate', along with an endless list of characteristics that one human being could hardly be expected to fulfil with any consistency, if ever at all. It is no longer good enough just to think of a suitable partner as being someone who has a good job, is pleasing to look at, can dodge a sabre-toothed tiger and might be worth giving a go.

What does it feel like to sign up to a dating site and what are the mathematics of love used by the algorithms? What has the dating game in the era of Big Data taught us about human beings?

The digital Hula Hoop

Dating sites offer profile-matching services to those for who spraying a handprint on a local car park wall just hasn't worked. Some of these sites are simply enormous. Badoo, with 245 million signed-up users, has a reach that covers most of the planet. It's clear that dating is big business. Match Group, whose portfolio of dating sites includes OkCupid, Tinder and Match.com, went public in November 2016 and currently boasts a market value of $2.3 billion.

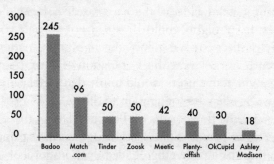

Having 'registered users' isn't the same as 'active users' (you would hope the service actually works and at least some of those who signed up abandon their online dating activities after finding the love of their life, even if only temporarily). According to Alexa Internet, a company which seeks to establish popularity ranking by tracking user activity through web browser toolbars, the most popular dating site is RocknRollDating.com, which matches people based on their musical preferences. You may be a supermodel and the other person may be a vet with a part-time modelling contract, and in the data-driven world of people-matching sites you are the ideal couple – but that Justin Bieber album lurking in your collection may just prove fatal to a fledgling relationship via RocknRollDating.com.

Second comes 'Right Stuff' a site for graduates and faculty members of Ivy League, Seven Sisters and fifty other universities and medical schools in the United States (clearly dusty academics have unrequited feelings, too). Interestingly, next on the Alexa list is ShakeMyWorld (a subsidiary of Match.com), located in Sweden. This may strike you as strange, given that surveys regularly declare that Swedish people are among the most uniformly beautiful people on the planet. You would have thought throwing a Hula Hoop on to a crowded dance floor on a Saturday night would be sufficient to nab you a suitable partner for life if looks were all that mattered. Apparently not. Also, with a population of 9.5m people, having 1.9m active users would imply that other nations than just Sweden are logging on to ShakeMyWorld.

Besides the behemoths of Internet dating, there are also the niche players. Specialist dating sites have popped up catering for a wide range of lovelorn hopefuls:

 TheBigandtheBeautiful: dating for plus-sized women who are looking for men who will love them just the way they are. 'We're all about just being honest,' says Whitney Thompson, President of the company and the first plus-size winner of America's Next Top Model. thebigandthebeautiful.com

VeggieDate: vegetarian dating for vegetarian singles. Vegans, lacto-vegetarians, ovo-vegetarians and their 'raw' variants are welcome, as well as pescatarians or those thinking of becoming a vegetarian. veggiedate.org

BeautifulPeople: what it says on the tin – ugly people need not apply. Not politically correct, but 'honest' according to director Greg Hodge. Who says beauty is only skin-deep? beautifulpeople.com

gk2gk: 'gk' as in 'geek': 'If you think you are a geek – you are a geek', according to the founder Spencer Koppel, a retired actuary. If you find yourself correcting yourself on minor infractions of fact on your chosen specialist subject, this might be the place for you. gk2gk.com

FarmersOnly: farmers live rurally, are less materialistic and get up very early – the sort of things city dwellers just don't understand. If the phrase 'if it ain't happenin' in the hen house, it's a happenin' in the barn' means anything to you, this could be the place to find love. farmersonly.com

These and many other services, such as Tinder, all work in a similar way. Only the method of delivery differs.

Let them eat cake...

As part of my research for this book (and with the full knowledge of my wife) I joined Match.com for a day, telling as much truth in the questionnaire as modesty would allow. As I was looking for data, not a date, I never toyed with anyone's affections, never 'winked' at anyone (the rather coy way of introducing yourself to someone). Here are my results, which include the people online, the views I received and how many people 'winked' at me, graphed on an hourly basis. This also enabled me to construct an index of success that I'll call my 'winks per view'.

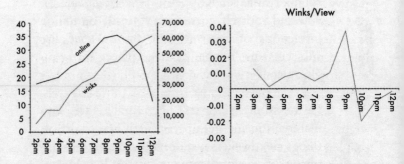

Pretty much from the get-go, I generated a lot of interest. Clearly, as a new man on the block, the novelty factor is enormous. There is a surprising amount of activity during working hours, but things really start to build in the evening. In terms of my 'winks per view', you can see from the graph that my popularity rises dramatically at 8 p.m., peaks at around 9 p.m. and then collapses dramatically. Not long afterwards, my new friends (sensibly) go to bed.

For a while, my 'winks per view' collapse was something of a mystery – until I realised that, on this particular day, my dramatic rise and fall coincided with the weekly showing of the UK's most-watched television programme, *The Great British Bake Off.* Clearly, the combination of seeking a partner and watching people bake cakes is a serious stimulant amongst my demographic, something I think is endearingly true and quite revealing.

Fortunately, I received only four emails. One of them was deleted, as it was clearly generated by a 'chatbot' – a program which sends out human-like conversations – none of which I replied to. Real communication between two actual people is a rich source of data from dating sites, however. You may not realise it, but once you are logged on to the site, not only is what you send being monitored, but also what you don't send: the number of keystrokes is being counted. According to research, those of a timorous nature will strike 387 keys before sending a 'Hey', but there is also another phenomenon that has only been recently recognised. There are a large number of people who appear to type fewer keystrokes than the length of text they send. The explanation is that about 20 per cent of correspondents use a 'copy and paste' strategy when responding to emails, which reduces the time and effort they have to put in before they meet someone. Somehow, that feels like a bad start to a relationship.

The mathematics of love, or 'That's very geometrically mean of you...'

Dating sites function with data supplied by users in the form of a questionnaire and a profile (a type of personal résumé). They need the data in a form a computer can eventually crunch and come to a statistical measure of your compatibility with another human being, otherwise you might as well take out a billboard advertisement with your name, preferences and the most flattering photograph you can find.

The algorithms behind dating sites and apps don't all work the same way. OkCupid asks users three questions: what they're looking for, what they want in their partner, and how important that trait is.

The 'importance' aspect of the questionnaire is very important to potential compatibility. At OkCupid there are four options for how important a question is to you and your prospective partner. The algorithm assigns a value to each answer, ranging from 0 to 250.

Level of Importance of Question	Point Value
Irrelevant	0
A little important	1
Somewhat important	10
Very important	250

At the risk of destroying my marriage, let's perform a simplified compatibility test on myself and my wife, Alice.

Stewart	Q1: Do you like watching Danish movies?	Q2: Do you like cooking?
Self	Yes	No
Wanted from partner	Yes	Yes
Importance	A little	Somewhat
Importance score	1	10

Importance score total = 10 + 1 = 11

Alice	Q1: Do you like watching Danish movies?	Q2: Do you like cooking?
Self	Yes	No
Wanted from partner	Yes	Yes
Importance	Very	Very
Importance score	250	250

Importance score total = 250 + 250 = 500

We now (somewhat scarily) have all the data needed to find out the compatibility score between Alice and me.

My score: Over the two questions, I answered 'a little important' and 'somewhat important', meaning these questions are worth 11 points. Alice answered, as I wanted, on both questions, meaning that she got 11/11. This makes Alice 100 per cent satisfactory to me.

Alice's score: Over the two questions, Alice answered 'very important' both times, meaning these questions are worth 500 points to her. I answered as Alice wanted on one of the questions, meaning I got $250/500 = 50$. Stewart is 50 per cent satisfactory to Alice.

Finding out that my wife is perfect, but I am only 50 per cent satisfactory, isn't really news to me, but it has been sufficient to sustain a thirty-year marriage so far. For data scientists, however, this isn't enough to test compatibility. They need to calculate the geometric mean of the two percentages. For two numbers, this means multiplying them together and taking the square root[1]:

$$\text{Compatibility score} = \sqrt{(100 \times 50)} = 71$$

So my wife and I are 71 per cent compatible according to this simple test, which is, frankly, a result I cling on to like a limpet to a rock! However, we only answered two questions, which isn't really enough information; OkCupid encourages users to answer at least a hundred questions.

Match.com doesn't seem to be as trusting. Its algorithm, a code named Synapse, notes not only your stated preferences, but what you actually do while on the site. The distance between them is called 'dissonance', which complicates the dating game markedly. If you say you like 'men under 26' but sneakily keep on looking at men over 30, then the Match.com algorithm will note

---

1 The geometric mean is the Nth root of the scores where N is number of numbers being multiplied together. It's a mathematical trick to get around the problem of the more familiar arithmetic mean which can be misleading e.g. imagine a couple of 0% and 100% compatibility. If we used the arithmetic mean they would be $(0 +100\%)/2 = 50\%$ compatible which clearly isn't the case because one is obviously a saint and the other a psychopath. Using the geometric mean gets rid of this problem i.e. $\sqrt{(0 \times 100\%)} = 0\%$

this and start offering up not only men over 30, but also people who are behaving like you. Essentially, Synapse is using the same techniques used by Amazon, Netflix and Spotify to find not only what you say you like, but also things that are close to what you like, and sell them to you. This way, you can be taken to an unexpected place where you might find love. It also relies on the dreaded feedback – the more you rate the suggestions, the more the algorithm can hone in on your real preferences. This is a powerful technique – maybe too powerful. In 1995, in the early days of Match.com when most people weren't online, founder Gary Kremen had all his friends, employees and even his own girlfriend sign up to boost numbers and give the site critical mass. His girlfriend met another man through Match.com and left him.

There is some danger in knowing how dating algorithms work: you might start gaming the system in order to get dates which aren't in your or anybody else's interests. However, that's not the same as taking some steps to remove some barriers to success. According to Amy Webb, founder of the Future Today Institute, you really should think about the content of your profile. No matter how smart you are and how much you want to project your accomplishments, ninety-seven words appears to be the maximum attention span most potential mates have. Be optimistic, which makes you approachable. Don't be tempted to respond immediately, leaving about twenty-four hours between communications. Most of all, play by your own rules and feel free to be as picky as you like.

You don't have to accept what a dating website throws at you. You could try to reverse engineer the situation and use dating sites in the same manner as a database, rather than allowing an algorithm to dictate who you should see. Ask your own questions about what you are looking for in a potential mate – even create your own scoring system – and scrape the data out of the profiles being presented to you and calculate your own compatibility score. When Amy Webb did this, she was proposed to after eighteen months, married a year later and had a baby a year after that. You can see her story on YouTube.

So far, we have only looked at dating websites – which, to be frank, seem to be for older people these days. The de-stigmatization of dating via a phone app has seen an explosion of usage for services such as Tinder in the past five years and it is apparently the route to love preferred by millenials (aged eighteen to thirty-four). As comedian Aziz Ansari says in his book *Modern Romance: An Investigation*:

> *Today, if you own a smartphone, you're carrying a 24-7 singles bar in your pocket.*

There is a mystery about how Tinder works. Sean Rad, chairman of Tinder, is somewhat reticent to discuss the algorithm behind it. However, the principle of swiping right for 'Like' and left for 'Don't like' when the app presents potential candidates appears to affect who is shown to you when added to your GPS location from your phone. There also appears to be a reward system at work: if you use the app more frequently, you get shown more and apparently better-looking people. If you receive a lot of right-swipes from other users, this appears to work in your favour. Tinder is akin

Dislike? (Close the book)

to an enormous voting system, a kind of perpetual referendum on attractiveness. The company has at least confirmed that its algorithm gives each user an 'Elo rating', a term borrowed from skill rankings in the chess world, which ranks desirability using more than facial recognition. If this sounds like a game of sorts, then you would be right. Tinder is supposed to appeal to the gaming generation and for those who find filling in long, tedious questionnaires just too much of a chore. Like gaming, you can get addicted to spending hours a day swiping right and left – which, in the world of online reinforcement, has the effect of improving your Elo score.

Initial contact from a potential mate will just be the beginning of the data trail. That person could be googling the hell out of you, looking for data and other clues before you actually meet. According to an online survey in 2011, 80 per cent of millennials said they did online research on a person before a first date.

The whole truth and nothing but the truth...

Online dating services are revealing more and more about our attitudes towards finding a mate. They're places where we are least inhibited, because our actions are private and our desires most free of social pressure. We can literally be ourselves, because there is no advantage to self-deception when it comes to finding a mate. It must be said that the amount of data is still

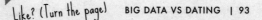

Like? (Turn the page) BIG DATA VS DATING | 93

relatively small – it doesn't cover an entire generation yet. However, insights are being drawn from the analysis of tens of millions of real expressions of preference. What it does offer is a snapshot of roughly where we are today. At worst, the conclusions emerging merely confirm our suspicion or at best confound some of our deeply held prejudices.

- Until the age of thirty, women say they like men who are slightly older than them
- At forty, women's tastes appear to hit a wall – they say they aren't much interested in men over forty
- Every age group of men consistently thinks women look their best in their early twenties
- For a woman, the pool of men attracted to them declines rapidly after the age of thirty
- Men are very influenced by the looks of a woman
- A woman's hair colour isn't very important to men
- The part of the political spectrum you occupy isn't a barrier to finding a date – people who think politics is important merely want to see that you care
- When OkCupid turned off the picture profiles of its members, their satisfaction with the subsequent dates didn't change – looks didn't matter
- Displaying a small imperfection (the 'pratfall effect') makes you appear more approachable and more attractive
- The four largest [self-described] racial groups on OkCupid – Asian, black, Latino and white – all get along about the same, i.e. no matter your race, compatibility score isn't affected

- Data taken from Match.com, OkCupid and DateHookup shows women display preferences that are more 'race-loyal' than men
- The two most under-appreciated groups on dating services such as Match.com, OkCupid and DateHookup are black men and women and Asian men
- Adding 'white' as part of your ethnicity increases your user rating on dating sites

I have lent heavily on some of the conclusions of Christian Rudder, owner of OkCupid, and his excellent and thoughtful book *Dataclysm*, to compile this list. However, the aggregated data gathered from the likes of OkCupid, Match.com and DateHookup is only registering initial reactions to images; they completely miss out the real-world dimension of sitting across from someone. This is still a fascinating glimpse into our true preferences using data as the basis of discussion, rather than anecdote and rhetoric. It fills the 'eerie numerical silence' surrounding our most contentious social issues.

Some guys have all the luck...

In his essay 'Metaphysics of Love', the philosopher Arthur Schopenhauer rationalised human attraction down to the cry of the unborn child. In it, he wrote: 'The real aim of the whole of love's romance, although the persons concerned are unconscious of the fact, is that a particular being may come into the world.'

In other words, we all start out as a twinkle in our parents' eyes, which requires, at the very least, two people to be in the same room with each other at the same time. How quaint and old-fashioned!

Even if computer-based dating is still too intimate and time-consuming for you, there is an alternative: miss out the whole anonymous dating scene completely and go straight to anonymous parenthood using profiling and data downloaded to your phone.

Of course, donors come from all walks of life, but according to the London Sperm Bank's (LSB) own website statistics, its regular monthly donors are dominated by IT managers and students – into which you can read whatever you want.

LSB's donors are classified by race, ethnicity, eye colour, skin tone, height, weight, highest academic qualification and occupation – all of which are searchable quantities on its database. You could, for instance, extract all Buddhists – even though it's not clear if a specific gene for Buddhism has been identified.

If this wasn't super-convenient enough, the whole process is available through an app. When a suitable donor becomes available, you will be sent a message.

A brief encounter

If YouTube statistics are to be believed, currently one of the most popular songs from the 1970s, with nine million hits, is 'Escape (The Piña Colada Song)' by Rupert Holmes. This touching ballad, in the narrative style, details the experience of a man who is bored with his marriage and places an advertisement in a lonely hearts page, then goes to meet a respondent who also has a preference for piña coladas. The man finds his wife waiting for him – she is the respondent. Despite the heartwarming message of the rekindling of a stale relationship through communication, this archetype for finding an extramarital mate is also a warning from the past: it can be a hazardous pastime, because you never know how or when you might be found out.

The Ashley Madison dating service's unique selling point was that it enabled extramarital affairs. If you wanted to lift the lid on specialist dating platforms and their security systems, then you would be hard-pressed to find a more interesting case study than the so-called Ashley Madison incident. The story began in July 2015, when a group of hacker activists calling itself 'The Impact Team' stole user data from the company. To add gasoline to the fire, Ashley Madison had a policy of not deleting personal information – users' names, addresses and credit card transactions – so the opportunity for public humiliation if such data was released was very real. In all, some 60 gigabytes of data was put into the public domain using the anonymous darkweb network Tor. The Impact Team were demanding the closure of Ashley Madison and its sister site, Established Men.

What was most fascinating was the analysis of the data undertaken by external parties. One fact that was apparent was that of the 4,000 easiest passwords to crack, the most popular were '123456' and 'password'. In all, 11 million passwords were eventually deciphered.

No matter how secure dating services are, nothing is 100 per cent secure.

Another strange revelation was that only 9,700 of the 5 million female accounts had ever replied to a message, compared to 5.9 million men. Annalee Newitz, the former editor-in-chief of tech website Gizmodo, concluded that there were a series of robotic algorithms talking to the men, persuading them they were talking to real women. It's even been suggested that the incident has been of benefit to humanity, if not mankind, as the female chatbots involved may have passed the Turing Test, a series of steps designed to discover if a machine's intelligent behaviour could be equivalent to or indistinguishable from that of a human. One thing it did prove is that a chatbot could be more intelligent than your average amorous male.

Big Data vs Dating

Good-looking people get the most attention, men like young women, older women like young men and physical attraction is a swiftly fading effect. None of these things are exactly news – they have been the subjects of just about every pop song going.

What we have learned from dating sites is that people seem to be liars, or rather they have a public face and a private face when it comes to expressing true preferences. There is nobody watching when you put a cross in a tick box on Match.com or similar. More accurately, there is nobody watching besides the people at Match.com, who know all your little secrets and have algorithms to work out if you're telling the truth or whether you really are the sort of person who dresses up in a clown outfit and stares into security cameras late at night.

As fascinating as this is, the question that should occupy our minds is whether we should know it at all. One of the most deeply worrying findings from the data gathered on dating services is the degree to which people will tailor their responses using a series of set pieces that they can cut and paste into a message box with the minimum of key strokes. Part of the success of any relationship is the desire to discover and keep on discovering, to unpick the Gordian knot of a personality by teasing out the entanglement over many years. Constructing a response of near perfection at the beginning seems like cheating (if it isn't actually cheating) to the point of laziness. Have we become so time-poor that responding to each other as unique individuals has become too onerous? Isn't the struggle part of the thing that binds us for the long-term? In reality, is the best dating site your train or bus journey, or your local coffee shop, if everyone could stop looking at their screens, swiping through their latest responses?

The owners of this vast amount of data – and they do own it, by the way – are also being faced with an interesting dilemma. They can see two things about humanity in a way we have never seen it before: those things that are immutable human laws and the development of shifting sociological trends in as near to real time as is possible, or even desirable. Sociologists, psychologists, marketers and governments would dearly love to get their hands on it. Together, they would all be a match made in heaven.

Today, if you own a smartphone, you're carrying a 24-7 singles bar in your pocket.

Big Data vs The Marathon

Should you chase the Womble when running a marathon?

As a sporting event, the marathon is a triumph of marketing over historical fact. There is no real time record of Pheidippides, Philippides, Thersippus of Erchia or Eucles, the name variously given to the runner who, about 2,500 years ago, is supposed to have slogged his way, in full armour, between the towns of Marathon and Athens to announce the victory of the heavily outnumbered Greeks over the Persians. In reality, there are no records of anyone having run the 26 miles and 385 yards (around 42km) of a marathon as we know it today, or even the more circuitous round-trip route proposed by the poet Robert Browning between Sparta to Athens (169 miles). We do know that the original Olympic race length in Athens in 1896, as reinvented by Michel Bréal, was 40,000 metres (about 25 miles). We also know that the extra mile and 385 yards was added so as not to inconvenience the family of Queen Alexandra in London in 1908 – her daughter and children could now watch the start of the race on the lawn of Windsor

Castle, then, at the end, Her Majesty would only have to raise her eyes from her programme to watch the contestants exhaustedly flop across the winning line directly in front her at the White City Stadium.

There is also a long and illustrious history of minimal preparation for the Olympic marathon. Never having run the event before seems to have been an advantage: Spiridon Louis (Athens, 1896), Delfo Cabrera (London, 1948), Emil Zátopek (Helsinki, 1952) and Alain Mimoun (Melbourne, 1956) were all winners who left their better-prepared competitors behind, and no doubt crossed the finishing line wondering what on earth all the fuss was about.

If you are wondering where all the women are in this clearly sexist account of Olympic marathoning, the answer is simple: they weren't allowed to run more than 200m until 1960, as it was deemed too distressing by the International Olympic Committee to see the 'gentler sex' perspiring. The first Olympic women's marathon took place at the 1984 Los Angeles games and was won by Joan Benoit of the USA, in a very respectable 2 hours, 24 minutes and 52 seconds (just 15 minutes and 31 seconds slower than the winner of the men's race). At the time of writing, the difference in world records between men and women stands at 12 minutes and 28 seconds. Kenya's Dennis Kipruto Kimetto scampered in after 2 hours, 2 minutes and 57 seconds in 2014, while Paula Radcliffe of Great Britain bobbed and weaved her way to the finishing line in 2003 after 2 hours, 15 minutes and 25 seconds.

Clearly, the methodology of running has improved since Spiridon Louis dragged himself across the line in 1896. Is there still something we can do using Big Data in our own training and fitness regime to improve our performance? Could we even find out if it is possible to compete in a marathon at all?

May the force be with you...

Some people start running as part of a fitness regime, others as part of a goal-orientated achievement, and both may be trying to lose weight. There is something oddly and uniquely dispiriting about running and thoughts of transformational weight loss, which comes from the world of physics. It doesn't matter how fast you run, you won't burn any more calories over a fixed distance. Here's the mathematics of running:

$$\text{Work} = \text{Force} \times \text{Distance}$$

The 'work' in this case is the energy (in calories, or 'kcal') needed to move your pallid carcass, whose weight (in kilograms) is the 'force' you have to apply over whatever 'distance' (in kilometres) you choose. It takes about one calorie to move one kilogram over one kilometre. If you weigh eighty-three kilos and you run five kilometres, here's how much you burn:

$$\text{Work} = 83 \times 5 = 415\text{kcals}$$

If you want to reduce the amount of calories you burn, shorten the distance (or lose some weight!). If you want to increase the calories burned, run further.

The interesting thing you should notice is that it is almost impossible to lose weight by exercise alone unless you are running massive distances every day, which isn't advisable. The reason is simple. Imagine you wanted to lose four kilograms. There are approximately 9kcals per gram of human fat. So that's $9 \times 4,000 = 36,000$kcals of excess calories to get rid of as fat. Rearranging our equation, that's $36,000/83 \approx 434$km. If you wanted to achieve that weight loss over two months, you'd have to run 7.2km every day, without fail and only if you were burning fat alone (which you won't).

In other words, it can't be (or it is highly unlikely) that you will achieve weight loss by exercise alone if you are training for a marathon. You should think of weight loss as a by-product of your marathon exercise regime rather than an aim. Think in terms of the benefits to your body shape. Having said that, be sure to invest in a belt before commencing a running programme to train for a marathon – pretty soon, your jeans will begin slipping over your hips as you trim down.

Sheer heart attack

If you are training for a marathon, you are going to need some data about yourself. You might like to invest in a fitness gadget, one that gives you the most information about heart rate, calories burned, GPS, footsteps and so on.

The usefulness of a fitness gadget is you can see your heart rate while you are exercising and while you are resting – it's data you're going to need to establish whether you should compete in a marathon, let alone complete one.

For people of a certain age, it's easy (and right), if taking up regular exercise after a long layoff, to become obsessed with the idea of having a heart attack. There is a good reason for this. According to clinical studies spread over many years and using over 20,000 patients, one of the best indicators of your future health is your resting heart rate. Investigators have also reported that, in both sexes and at all ages, cardiovascular and coronary mortality rates rise progressively as your resting heart rates rises. The effect is more marked in men than in women.

According to the (ironically named) Mayo Clinic, a good rule of thumb for a safe resting heart rate is 60–100 beats per minute. When exercising, your safe heart rate is calculated using your age:

$$\text{Safe heart rate (SHR) in beats per minute (bpm)} = 220 - \text{Age}$$

For a 55-year-old male, that's 220 − 55 = 165bpm. Of course, this could give you false comfort. Some people have questioned the scientific accuracy of sports devices (this is why you should get checked out by a doctor first) but it's a good rule of thumb. At least it will stop you thinking along the lines of, 'Hey! 185bpm – let's see how far I can PUSH this!' Besides, the idea is to just raise your heart rate for a while, rather than inducing it to imitate something that could pump fuel around a jet plane.

Now you have your resting heart rate and your SHR, you can calculate your target heart rate zone; the range your heart rate should stay whilst exercising. A vigorous exercise regime would target a heart rate of 70 to 85 per cent of the difference between your resting heart rate and your SHR. You would calculate it like this:

1. Subtract your age from 220 to get your maximum heart rate
2. Measure your resting heart rate using your fitness gadget
3. Calculate your heart rate reserve (HRR) by subtracting your resting heart rate from your maximum heart rate
4. Multiply your HRR by 0.7 (70 per cent). Add your resting heart rate to this number
5. Multiply your HRR by 0.85 (85 per cent). Add your resting heart rate to this number
6. These two numbers are your training zone heart rate. Your heart rate during exercise should be between these two numbers

For example, say your age is 55 and you want to figure out your target training heart rate zone. We already know how to calculate your maximum heart rate. In the example above, it's 165. Next, calculate your HRR by subtracting your resting heart rate of, say, 60 beats per minute from 165. Your HRR is 105. Multiply 105 by 0.7 to get 73.5, then add your resting heart rate of 60 to get 133.5bpm. Now multiply 105 by 0.85 to get 89.25, then add your resting heart rate of 60 to get 149.25bpm. Your target heart rate zone for your training schedule should be between 133.5 and 149.25 beats per minute.

Using fitness gadgets, you may notice (especially at the beginning of a training regime) that your heart rate

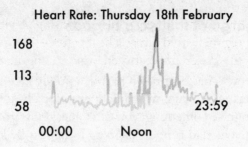

Heart Rate: Thursday 18th February

168

113

58   23:59

00:00   Noon

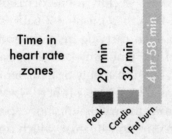

Time in heart rate zones

29 min — Peak

32 min — Cardio

4 hr 58 min — Fat burn

will stay elevated for several hours after a run. This Fitbit graph shows my heart rate. The dark blip at the top is me on a treadmill. The lower middle part of the trace indicates when my heart rate is in the 'fat burn' zone. The bottom section is when my heart is in, or retreats to, its resting rate. Look at the way the 'fat burn' zone lasts for several hours after exercising. It means that, even though I was sitting down, my body was experiencing something that was more like walking vigorously. As you get fitter, the effect attenuates — but using data like this can be extremely helpful in gauging how much exercise you are taking each day. On this occasion, I spent nearly five hours in the 'fat burn zone', most of it while watching television. Without data, I couldn't have known that and adjusted my training schedule to suit my goals.

The charge of the Lycra brigade

Anyone who has ever taken part in a public race such as the fantastic parkrun, which happens globally at 9 a.m., will recognise something: you can't guess how good any of the Lycra-clad runners is going to be just by looking at them. My personal nemesis became a small, slightly boxy-shaped woman who would habitually plod past me on the second lap every week. On one occasion, I was sequentially passed by a man pushing a baby, a lady in her seventies and a man carrying his girlfriend on his back. I haven't had the ignominy of being passed by another species yet, but it may only be a matter of time. Clearly, there is something going on here – something that is concealed from the naked eye, but separates us and will impinge on our ultimate goal, which is to run a marathon.

That 'something' is an inescapable concept called the VO2 max. Put simply, a VO2 max is the maximum rate of oxygen consumption (and therefore, energy creation) the body can perform. It's a measure of aerobic fitness. Since your personal running limitation will be set by how quickly you can consume oxygen (create energy), it's going to be a pretty important vital statistic. It's actually measured in millilitres per minute. Most commonly, it is expressed as the amount of oxygen absorbed per kilogram to allow comparison across body types – larger people tend to use more oxygen per minute. But the main thing to remember is the higher the number, the better you are at energy conversion and the more endurance you will have.

The bad news is that a lot of your personal VO2 max number is genetic. Some people respond to training and can increase their VO2 max by about 10 per cent, but we aren't all capable of reaching the VO2 max of an elite athlete unless we are already elite athletes. You can be all you can be, but you can't be all of what someone else is. Studies show your VO2 max decreases with age and is gender-dependent – no amount of training could have allowed Paula Radcliffe to bridge the 12 minutes and 28 seconds between her and Dennis Kipruto Kimetto. VO2 max levels are generally 40 to 60 per cent higher in men than in women, mainly due to the differences in body weight and lean body mass.

To find out your VO2 max, you could put yourself through the hassle of going to a medical facility to be wired-up while on a treadmill. Fortunately, there is a more leisurely way to get the data, based on the output of your fitness gadget. You can input results into one of the equations derived from numerous studies of many athletes to calculate your VO2 max. Such results are in the right ballpark and certainly good enough for amateur fitness enthusiasts.

One of the equations you can use is the intimidatingly named Uth-Sørensen-Overgaard-Pedersen estimation. Just pronouncing the name will probably have most people reaching for an oxygen mask, but it looks like this:

$$VO2\ max \approx 15.3 \times HRmax/HRmin$$

HRmax is your maximum heart rate and HRmin is your resting heart rate (which become abundantly clear if you have a Fitbit, Garmin or suchlike). My VO2 max ≈ 15.3 × 165/60, which is 42.1 on a good day. There are various equations like this, but they all come out pretty much the same. My VO2 max of 42 is, according to the league tables, good bordering on excellent for my age. I very much doubt I will get much further than this without allowing my maximum heart rate to drift up (not advisable) or training so much that my resting heart rate plummeted (unlikely). The chances of me ever having a VO2 max above 45 are pretty much minimal, so I should just relax about it.

To find out where you are for your age and gender, there are helpful tables around such as these:

Men	Age (Years)					
Rating	18-25	26-35	36-45	46-55	56-65	65+
Excellent	>60	>56	>51	>45	>41	>37
Good	52-60	49-56	43-51	39-45	36-41	33-37
Above Average	47-51	43-48	39-42	36-38	32-35	29-32
Average	42-46	40-42	35-38	32-35	30-31	26-28
Below Average	37-41	35-39	31-34	29-31	26-29	22-25
Poor	30-36	30-34	26-30	25-28	22-25	20-21
Very Poor	<30	<30	<26	<25	<22	<20

Women	Age (Years)					
Rating	18-25	26-35	36-45	46-55	56-65	65+
Excellent	>56	>52	>45	>40	>37	>32
Good	47-56	45-52	38-45	34-40	32-37	28-32
Above Average	42-46	39-44	34-37	31-33	28-31	25-27
Average	38-41	35-38	31-33	28-30	25-27	22-24
Below Average	33-37	31-34	27-30	25-27	22-24	19-21
Poor	28-32	26-30	22-26	20-24	18-21	17-18
Very Poor	<28	<26	<22	<20	<18	<17

Elite athletes go way beyond this. They are, frankly, not really of this world. At his peak, five-time Tour de France winner Miguel Indurain is reported to have had a VO2 max of 88, while cross-country skier Bjørn Dæhlie's was measured at 96. The drug erythropoietin (EPO) can boost VO2 max by a significant amount in both humans and other mammals, explaining its attractiveness and illegality in sports such as cycling. Siberian huskies running in the Iditarod Trail Sled Dog Race have VO2 max values as high as 240, which is some three times the VO2 max of an elite human athlete. Do not challenge a Siberian husky to a race, no matter how drunk you are.

Diet another day

To complete or even take part in a marathon, you have to make other changes to your life. One of the most important is to stop drinking. You certainly shouldn't follow the example of either Thomas Hicks or Charles Hefferton. Hicks, an English-born American, was given so much strychnine (as a stimulant) and brandy by his team entourage during the 1904 Olympic marathon in St Louis that he swore never to race again. Charles Hefferton of South Africa foolishly accepted a glass of champagne from a spectator at the 1908 London Games when leading an Italian confectioner called Dorando Pietri by four minutes. Hefferton went on to get stomach cramps, and lost his lead. Film of the event shows Pietri, with his trademark knotted handkerchief on his head, entering the stadium first in an advanced state of collapse, turning the wrong way and then taking a tortuously wayward route to the winning line, surrounded by policemen and officials.

There are no two ways about it. All the fancy diets in the world really boil down to this: stop drinking, remove as much sugar from your diet as you can tolerate, forego pasta, bread and potatoes (unless you are specifically fuelling up for a race). Eat vegetables. Everything else is marketing and packaging, designed to get you to the same conclusion. Mainly, stop drinking. There have been wide-ranging studies showing that moderate alcohol intake (one drink for women and two drinks for men) can have some benefits, such as the polyphenols in red wine, which are thought to be protective against cancer. In general, though, booze will:

- Reduce the performance of middle and long-distance runners
- Impair the repair of injuries
- Impair your ability to recover from training

There is another reason, which is down to mathematics. A bottle of wine contains about 600kcals. Rearranging the equation above, this equates to just over seven kilometres of running for me (600/83 ≈ 7.2). That's a lot of running just for a night out, or indeed, a night in.

What's worse is that alcohol calories aren't like the calories in a lettuce leaf. Alcohol is metabolised in the body as fat. This will seriously impact your ability to lose weight and get you in shape for a marathon. Each day, an average male burns about 2,500kcals just by being alive. To break even, you'd have to consume less than 1,900kcals a day if you drink a bottle of wine. If you are a woman it's even worse: women are recommended 2,000kcals a day, so they can only have 1,400kcals.

A bottle of wine is way over the approximately two bottles of wine a week recommended by doctors as a low-risk alcohol intake, but as we all know, it's easy to fall into bad habits. Needless to say, drinking and running a marathon really don't mix, mainly because the zigzagging you'll be doing will only add to the distance you cover, increasing your time.

Just another brick in the wall

If I wanted to run a marathon using our first equation (here, using the metric length of the course at 42.2km), it's easy to calculate how many calories I would need:

$$\text{Work} = \text{Force} \times \text{Distance}$$
$$83 \times 42.2 \approx 3{,}502\text{kcals}$$

The question then becomes one of having enough energy stored in the muscles to satisfy the needs for a marathon. If I don't, I will hit what is commonly called 'the wall': legs suddenly become independent of the body and a runner resembles a Lycra-clad Elvis staggering down The Mall at the end of the London Marathon. Remember: the cameras will focus on you. It's best avoided.

The problem is that the body is very selective about where it stores energy. If it isn't at the source where it's needed, then it won't get it from somewhere else. There is some generally available in your liver, but the glycogen energy has to be stored in your legs and buttocks to enable you to complete a marathon.

A runner can preload with carbohydrates in the twelve to thirty-six hours before a race, but that doesn't guarantee not hitting the wall. Whether it is hit or not depends also on the VO2 max. This means the runner needs a target time, to set off and maintain the right pace, to stick to it and not deviate from it.

The calculation for your own personal wall is fiendishly complex, because it depends on many variables, including sex, age, weight, VO2 max and the time in which you want to complete the marathon. Fortunately, help is at hand. Data from thousands of runners over many decades has allowed scientists, most notably Benjamin Rapoport, to model and personalise a runner's expectations and allow them to plan to compete in (and most importantly, finish) a marathon. All you have to do is input your vital statistics and out comes the answer.

There are websites designed to help you calculate and set your expectations for a marathon. Here are mine for a target time of four hours, taken from endurancecalculator.com:

Basic Information	
Estimated Maximal Oxygen Uptake (VO2 max) (in mL oxygen per kg body mass per minute)	42
Total Energy Required to Run a Marathon (in kcal or food calories)	3,502

Conservative Best Marathon Performance (Normal Glycogen Levels)	
Time	4:35:02
Mile Pace	10:29
Kilometre Pace	6:31
Carbohydrate Loading Required (kcal)	1,837

Aggressive Best Marathon Performance (Maximal Glycogen Levels)	
Time	3:08:39
Mile Pace	7:11
Kilometre Pace	4:28
Carbohydrate Loading Required (kcal)	3,326

Target Marathon Performance	
Time	4:00:00
Mile Pace	9:09
Kilometre Pace	5:41
Carbohydrate Loading Required (kcal)	2,267
Minimal Fueling Required During Race (kcal)	0

According to this data, it turns out that a marathon is in my range, without refuelling, if I run at a pace of 5 minutes and 41 seconds per kilometre and load up with 2,267kcals before the race. However, this is a problem. Right now, I can't do 10 kilometres at that pace, let alone just over 42. Either there is a lot of training to do, or I may never be able to do it at all. I could slow it down, using the 'conservative' assumptions and do it in 4 hours and 35 minutes at a much more sedate 6 minutes and 31 seconds per kilometre (my comfortable pace is 5 minutes and 28 seconds outside and a much more aggressive 4 minutes and 48 seconds per kilometre on a treadmill). The only problem is that Dennis Kipruto Kimetto would be 2 hours and 32 minutes ahead of me.

Big Data vs The Marathon

There is a stark difference between men and women as they attempt to set faster records. Men's and women's marathon times since the 1950s show, at the current rate of progress, that men won't even approach the two-hour barrier until 2033, while, on the current evidence, women's progress may have plateaued. Even allowing for Paula Radcliffe's astonishingly outsized improvement (which some say may never be bettered), women aren't due to cross the marathon line in less than two hours until the year 2120. Using a data-driven approach, combined with advanced training techniques derived from Big Data studies, may yet see these milestones surpassed before the extrapolated dates.

Some people are very mistrustful of the sports gadget-derived data approach to fitness, seeing it as a battle between technology and listening to yourself – and with good cause. Health is dependent on more things than the limited metrics you can get out of a fitness wristband. Make no mistake, running a marathon is a dangerous thing to do. A 2017 study by Yale University discovered that 82 per cent of the runners taking part in the Hartford Marathon crossed the line with the early stages of acute kidney failure. The causes could include dehydration, a rise in core body temperature and a drop in blood flowing to the kidneys.

There are several really good reasons to know your body, your VO2 max and how Big Data can help us with fitness and training, even if it is just the beginning of the journey. Not least of this is that no matter how much you train, for you life may have just dealt the hand that means you really shouldn't try pushing yourself beyond a certain point. Know your limit and stick to it.

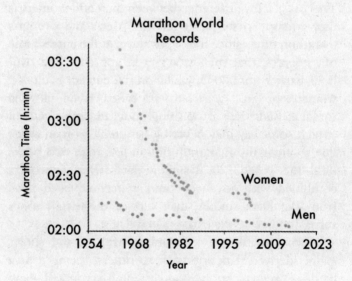

This may sound somewhat defeatist, but there is one very good reason to pay attention. Anyone who has ever attempted a recreational running event in the UK will be familiar with a syndrome called 'Get the Womble'. It goes like this. At some point during a race, you'll be passed by someone wearing a character outfit in the shape of say, the fictional Wombles. This affront to your dignity will lead you to abandon your carefully crafted race strategy and start chasing the animated character. Not very long afterwards, you'll find yourself sitting on the edge of the road, exhausted and forced to abandon the race as the Womble disappears into the distance. What you didn't realise is that inside the Womble costume was an elite athlete who was challenged to run the race a week ago by his mates in the pub while he was sipping a lightly carbonated mineral water. Knowing your VO2 max and your body may help you to avoid such folly. Let the Womble go.

You should also get the sense that not all of your failures (or even successes) at aerobic sports (especially running) are your fault, or even down to your hard work. The latest research has even suggested our psychological perceptions of, and response to, 'effort' may be as important to us as anything physical. Better brain training may be our next leap forwards. Knowing some mathematics, gathering your personal data and combining it with the models derived from studying thousands of runners may actually help you through the early difficult months and encourage you to persist with your training to run a marathon even in your darkest moments. Hell, it might even become enjoyable one day…

Big Data vs The Bookies

Should you bet against the robots?

For a Texas horse-trader, William Lee Bergstrom chose the right place when he walked into suitably named Binion's Horseshoe Casino in Las Vegas in 1980 with two suitcases. He had $777,000 in cash in one suitcase and nothing in the other: that one was for his winnings. The casino had guaranteed to match any bet that came through the door, as long as it was the punter's first bet. Bergstrom wanted to play a game of craps – in other words, rolling dice. Two rolls later, the eponymous Benny Binion was gallant enough to help Bergstrom stack his winnings into the empty suitcase, which Bergstrom took and disappeared for a few years. He later returned and repeated the feat a second time, pocketing a mere $655,000 in total after more successful rolls. There was also a seat at the Willy Nelson show for him and his mother to round off what must have been a satisfying evening. Bergstrom's luck finally ran out in early 1984, when he lost one million dollars on a bad bet.

He probably didn't realise it at the time, but Bergstrom was both at the beginning and at the end of something. He was operating at a time of relative innocence. It was a time when you could sit opposite a dealer or opponent and you looked them in the eye, making it a game of psychology as well as a game of chance. Pretty soon, the Internet would emerge and all that would change. Online poker saw both the dealer and opponent retreating behind screens. One of the first things the Internet found a use for, besides esoteric avant-garde poetry sites, dating and pornography, was gambling. It was like humanity's vices had gathered around the new kid on the block in order to corrupt it – helped by changes in gambling laws.

First out of the blocks was Antigua and Barbuda, which passed the Free Trade and Processing Zone Act, giving the small Caribbean nation the right to grant licences to those businesses interested in starting online casinos – which they promptly did. Not to be outdone, the Kahnawake Gaming Commission in Canada (actually a Mohawk First Nations commission) issued online gaming licences soon after. The very first gambling websites started to appear in the mid-1990s. From obscurity to $830m of revenues in 1998 took just a few years. Combined with the technology from companies such as Microgaming and CryptoLogic to enable secure transactions, online betting had arrived.

How do you gain access to this world and how does it work? More to the point, how is Big Data aiding online bookies and punters? Is it really possible to beat the bookies using it?

My future's so bright I have to wear shades

The size of the online gambling market (split between betting, bingo, casinos, the lottery and poker) is dazzling and is growing brighter by the year. With an 11 per cent annual growth rate, the total marketplace for global online gambling will head towards $70bn by 2020, according to some estimates. Given the uncertainties surrounding accurate data on the number of gambling accounts, this is almost certainly a gross underestimate.

Of the eighty nations that now have legalised online gambling, the countries of Europe are by far the biggest market, with an annual growth rate of 15 per cent. In countries such as Singapore, the government has made all forms of gambling illegal, while operators of gambling sites working illegally are routinely arrested in China and South Korea. This didn't put off some of the more determined Chinese bettors, who merely relocated to the Philippines. In November 2016, the Philippine authorities arrested 1,318 Chinese online gambling workers in a single raid. The USA has an unusually illiberal, and even inexplicable, approach to online gambling. Only three states currently allow it (Delaware, Nevada and New Jersey, but with more set to follow). In

Africa, similar to Asia, there are no plans for legalizing online gambling any time soon. In South Africa, despite online casinos being illegal, they still operate and it is left to the banks who operate the payment systems and the casino operators themselves to police gamblers' activities. In practice, a blind eye is turned to the whole thing.

One country that stands out in the new data-driven world of gambling is the United Kingdom. Unlike other nations, it allows gamblers from around the world to log on to its online betting sites – between 2014 and 2016, the number of active accounts in the UK grew from nearly 19 million to over 24 million, which is remarkable for a population of 65 million people. A 2016 report by the UK's Gambling Commission described how traditional betting shops and bingo halls are continuing to close every year, while online gambling was growing at a rate of 33 per cent, outstripping the combined takings from traditional betting and casinos. Even after handing back any winnings, online betting raked in £4.8bn (about $5.8bn) in 2016 for its operators, on a total of £41.6bn ($49.9bn) of bets. The online gambling market is divided into betting, bingo, casinos and 'pool betting', which is by far the largest segment of the online betting market. Football-related betting dominates, bringing in nearly £600m ($717m) for the bookies – almost double the amount taken from horse racing.

Perhaps surprisingly, it seems that gamblers prefer to access online betting via desktop computers. This is apparently because of the bigger screen size (it's possible to see your account funds dwindle with greater resolution) – with 60 per cent of UK gamblers saying they use their laptop. Still, the main service providers

World Map of Online Gambling

U.K. - 2016 profit = £4.8bn

U.S. (except Delaware, Nevada, New Jersey) - Still illegal

China - Still illegal

South Korea - Still illegal

Africa - Still illegal

Singapore - Still illegal

continue to push mobile devices, which are most popular amongst men (women prefer their tablets). It's difficult to watch a sporting event without seeing advertisements for in-play betting (betting that happens during a sporting event) from the likes of William Hill, Paddy Power, Bet365, Net Entertainment, BetFair, Unibet or 888 Holdings. Most of the time, the actual sportspeople appear to be animated hoardings for the same companies which allow betting on them, thus completing an almost dazzlingly circular argument.

Getting started in online gambling is seductively simple. All you need is a debit or credit card and a self-declaration that you are over eighteen years old. If you would like to use an alias such as 'SharkFish', or even to honour the memory of William Lee Bergstrom by using his name, then you can. You can set limits of what you can deposit within one, seven or thirty days. However, these can be amended, so offer little, if any, deterrent for someone who gets hooked and is tempted into the martingale system (the process of doubling your bets each time you lose until you win, which can see bettors betting large sums to win back their original small sum). Most importantly, you may be enticed with a joining bonus that claims to 'double' your initial deposit. Read the small print, though. Often, before you can make a withdrawal, you may have had to bet several times the value of your original stake. For instance, if you deposit £100, you may be required to have staked £600 within 90 days of signing up to become eligible for the £100 bonus stake. It's more than likely you will have lost more than £100 by the time you reach the £600 level, because the longer you play, the more likely it is you will lose (as we shall see).

Then there are the other companies who exist for the sole purpose of sucking in the unsuspecting, bleeding them dry, taking their personal information and running off with their funds and identities. Dead giveaways for such sites are poor spelling and odds that appear too good to be true. However, even the most reputable sites are prone to glitches. On 28 December 2011 in the Christmas Hurdle at Dublin's Leopardstown racecourse, £800,000 ($960,000) was placed at the astronomical odds of 28-1 on a horse called 'Voler La Vedette' creating winnings of £23m ($28m) for 200 punters.

More surprising are the circumstances. At the time the money was put up, Voler La Vedette was still running – or more to the point had actually passed the post – in first place. Needless to say, Betfair blamed a rogue bot (a malfunctioning automated bet-placement program) for the glitch and reneged on the bets (besides some goodwill payments), leading to some very, very angry punters. The incident instantly wiped £40m off the value of the company. But it only goes to show that even the best and most reputable software in the area of online betting is open to mistakes from time to time.

The wizard of odds

'Sports investing' is what some people call betting on sporting events, which carries the implication of equivalence to, say, playing the stock market. In two important respects, this is right. Beating the market as an investor can happen in three ways:

1. You spot an investment is mispriced
2. You have a piece of information a price doesn't reflect
3. You are just lucky

For a gambler, the equivalents would be:

1. You spot the odds on offer are too generous
2. You have a piece of information you believe is not reflected in the odds
3. You are just lucky

There, the similarities end. Buying a diversified portfolio of stocks and bonds rarely results in you losing everything in one day – more often than not, you are left with something in your portfolio. In the gambling game, you are guaranteed to lose: it's built into the system.

You may have heard someone saying phrases such as, 'The chances of that happening are five to one' and so on. It's important to understand what odds and relative probabilities are and how they relate to each other. The odds of 'a to b' represent a relative probability of $b/(a + b)$, e.g. 4-1 (four to one) is $1/(4 + 1) = 1/5 = 0.2$ (or 20 per cent).[2]

Think about a football match. There are three possible results: a home win, a draw, or an away win. You might expect the following odds to be encountered to represent the true chance of each of the three possible results:

---

2 A *relative probability* of x represents *odds* of $(1 - x)/x$, e.g. 0.2 is $(1 - 0.2)/0.2 = 0.8/0.2 = 4/1$ (4-1, 4 to 1).

Home win: Evens (1-1)

Draw: 2-1

Away win: 5-1

These odds can be represented as relative probabilities (or percentages, by multiplying by 100) as follows:

Home win: Evens (or 1-1) corresponds to a relative probability of ½ = 50 per cent

Draw: 2-1 corresponds to a relative probability of ⅓ = 33⅓ per cent

Away Win: 5-1 corresponds to a relative probability of ⅙ = 16⅔ per cent

Adding the percentages together gives a total 'book' of 100 per cent (called a 'fair book'). If you applied these odds to this situation and let three friends bet £50, £33.33 and £16.67 against each other to win £100, your 'book' of business would look like this in a spreadsheet:

Result	Bet	For	Against	Relative Probability	Win
Home	50.00	1	1	50.00%	100
Draw	33.33	2	1	33.33%	100
Away	16.67	5	1	16.67%	100
Total	100.00			100.00%	

Result	Bet	For	Against	Relative Probability	Win
Home	50	1	1	=D2/(C2+D2)	=(B2*C2/D2)+B2
Draw	33.33	2	1	=D3/(C3+D3)	=(B3*C3/D4)+B3
Away	16.67	5	1	=D4/(C4+D4)	=(B4*C4/D4)+B4
Total	=SUM(B2:B4)				=SUM(E2:E4)

As a bookmaker in this situation, you are effectively performing a social function of facilitating an amusing pastime between old school chums. Someone is going to walk away with all of the £100 that has been bet in total. It should give you a warm fuzzy feeling inside, if you weren't in the betting business.

As a bookmaker, you don't get to drive around in a powder-blue Rolls-Royce by being fair or acting as a social facilitator. So you tweak the odds (just a little) and reduce them to guarantee you make a profit no matter what happens.

The simplest way to do this, and guarantee a profit, is to proportionally decrease the odds. Using the example we have been working with so far, the following, slightly reduced odds, are in the same proportion with regard to their relative probabilities (3:2:1) but a remarkable thing happens:

Home win: 4-6

Draw: 6-4

Away win: 4-1

4-6 corresponds to a relative probability
of $6/10 = \frac{3}{5} = 60$ per cent

6-4 corresponds to a relative probability
of $4/10 = \frac{2}{5} = 40$ per cent

4-1 corresponds to a relative
probability of $\frac{1}{5} = 20$ per cent

By adding these percentages together, the 'book' now adds up 120 per cent. At first, this sounds somewhat strange. The amount by which the actual 'book' exceeds 100 per cent is known as the 'overround', 'bookmaker margin', 'vigorish' or 'vig'. It represents the bookmaker's expected profit. Thus, in an ideal situation, if the bookmaker accepts £120 in bets at his own quoted odds in the correct proportion, he will pay out only £100 (including returned stakes), no matter what the actual result of the match:

A stake of £60 at 4-6 returns £100
(exactly) for a home win.

A stake of £40 at 6-4 returns £100
(exactly) for a drawn match.

A stake of £20 at 4-1 returns £100
(exactly) for an away win.

Result	Bet	For	Against	Relative Probability	Win
Home	60.00	4	6	60.00%	100
Draw	40.00	6	4	40.00%	100
Away	20.00	4	1	20.00%	100
Total	120.00			120.00%	

Total stakes received are £120 and a maximum payout of £100 irrespective of the result. This £20 profit represents a 16⅔ per cent profit on all the money bet.

The built-in margin in favour of the house (which can range from a few percentage points to tens of percentage points) is why Las Vegas casinos can offer free drinks, meals and shows to punters. If you are spending $1,000 and have a 20 per cent vigorish, the house can afford to spend $100 on you and still walk away with $100. If you are a high-roller (or so-called 'whale') spending $250,000 in a weekend, the economies of scale kick in. It's unlikely you could spend $25,000 on comps like jet travel, limousines, food and Willy Nelson shows. The margins on whales are so much better and the numbers larger.

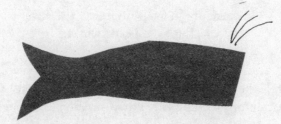

For gamblers, there is some other slightly bad news: to win money over the long-term, you have to win more than 50 per cent of the time. Numerically, if you go on a winning streak, the closest you will ever get is 11-10 (eleven wins to ten losses). In twenty-one events you'd have to win eleven and expect to lose ten, or win 52.4 per cent of the time, just to break even. If you ever hear of anyone with a win ratio of 55 per cent, then they are exceptional or pulling your leg.

You can make rational judgments about the odds you are being offered using a concept called the 'expected value' (EV), which is the amount a player can expect to win or lose if they placed a bet with the same odds over and over again. It's the average you might win or lose.

The equation for EV looks like this:

EV = (probability of winning × amount won per bet)−(probability of losing × amount lost per bet)

Let's imagine a football game between Chelsea FC and Hartlepool United. You want to bet on Hartlepool United winning (i.e. not losing or drawing) and you have £10 to put on as your bet. Here are your odds:

Chelsea FC to win: 5-4, which equals 1.25 to 1

Hartlepool United to win: 13-1

Draw: 6-1

To find the probabilities from the odds, take the reciprocal (turn them upside down and divide through):

Chelsea win = 4-5 = 80 per cent

Hartlepool to win = 1-13 = 7.69 per cent

Draw = 1-6 = 16.67 per cent

We want to know the probability of Hartlepool United not winning, which is the sum of Chelsea FC winning or drawing: 80 per cent + 16.67 per cent = 96.67 per cent.

Using the odds on Hartlepool United to win, we know that our potential winnings are 13 × £10, minus our original stake of £10 = £120. We now have all the information we need to calculate the EV:

EV = (probability of winning × amount won per bet)−(probability of losing × amount lost per bet)

$$EV = (0.0769 \times 120) - (0.9667 \times 10)$$

$$= 9.23 - 9.67$$

$$= -0.44$$

Which means, on average, you will lose 44 pence with the odds on offer.

One of the similarities between gambling and investing is based on finding an anomaly or an 'edge' and exploiting it. Casino floors, racetracks and online betting forums are like the trading floor of any international stock exchange; gamblers and traders are able to search around and compare the odds (prices). For instance, in the world of football betting, biases arise from time to time in the odds, which makes betting on a match to end with a draw more attractive than the win/lose punt. You could create a spreadsheet to do the EV calculations for you, which may help in assessing the odds to maybe beat the online bookies.

	Chelsea	Hartlepool	Drawn	Not Hartlepool
Odds	1.25	13	6	
Probability	80.00%	7.69%	16.67%	96.97%

Probability of Winning	Amount Won	Probability of Losing	Amount Lost	EV
7.7%	120	96.7%	10	−0.44

	Chelsea	Hartlepool	Drawn	Not Hartlepool
Odds	=5/4	13	6	
Probability	=1/B2	=1/C2	=1/D2	=(D3+B3)

Probability of Winning	Amount Won	Probability of Losing	Amount Lost	EV
=C3	=(C2*D6)-D6	=E3	10	=(A6*B6)-(C6*D6)

Now let's imagine you go to another online betting site which is offering 4-3 (1.3333 to 1) on a Chelsea FC win, 15-1 for a Hartlepool United win and 6-1 for a draw. Let's run it through our EV calculator. On average, probability would say that you would actually make 17 pence with those odds – a profit, which is a rare beast in the world of gambling. You might like to switch your bet from the loss-making site to the money-making site.

Alternatively, you could combine the odds from two or three sites to create a bet on all three possible results with the maximum possible EV. It won't guarantee that you'll win, but it will at least create a bet which, on average, sees you lose everything very slowly. It may be your best bet to beat the bookies.

	Chelsea	Hartlepool	Drawn	Not Hartlepool
Odds	1.33333	15	6	
Probability	75.00%	6.67%	16.67%	91.67%

Probability of Winning	Amount Won	Probability of Losing	Amount Lost	EV
6.7%	140	91.7%	10	0.17

And pigs might fly...

Making a judgment on whether those odds are 'right' or not is critical to beating the bookies and finding your edge. At the beginning of this chapter, the other way we identified gaining an advantage was spotting something that someone else hasn't noticed. This is the natural habitat of Big Data, so when Big Data and advanced analytics came blinking into the light, it got the attention of both the gambling industry and punters straight away.

The kind of mathematical techniques people have been using is known as 'regression analysis', as well as looking for statistical anomalies and machine learning.

Systems called Bayesian networks link seemingly unconnected factors to predict the future. Data on team performance, fatigue and motivation have been fed into programs using Bayesian networks to predict English Premier League football matches and beat the bookies' published odds. Sometimes, the output from statistical analysis can make uncomfortable reading for some supporters. Statisticians Håvard Rue and Øyvind Salvesen of the Norwegian University of Science and Technology used a Bayesian model to show that Arsenal football club 'shouldn't' have won the 1997–98 English Premier League title.

There is a simple way of illustrating how you can collect data to seek an edge in the world of online gambling. Take, for instance, the rivalry between Argentinian soccer teams Boca Juniors and River Plate – it's one of the oldest and fiercest in sport. They first met on 24 August 1913 and nothing has stopped them from meeting at least once a year in either the league or cup competitions, or both. Neither two world wars nor the Falklands conflict has been deemed sufficiently important enough to get in the way of their fixture. The meeting has even spawned its own description – the Superclásico. If hatred is too strong a word to describe the relationship between the two sets of supporters, then mere rivalry is certainly insufficient. At times, the match resembles a twenty-two-man kickboxing bout into which someone has incongruously introduced a football. If, at some point, riot shields aren't evident on the field of play, supporters begin to wonder if they are at the right game. It is usual that both poultry and livestock attend the game. By now, you get the picture and you also understand why this fixture might be the subject of intense betting.

If you downloaded the scores from all the matches going back to 1913 and created a graphical representation of them, they would look like this (see Graph 1). A data scientist could put a trend line through the long-term data using a regression model to try and predict the next probable data point in the series. That scientist might notice that, since 1990, there has been a distinct divergence in the results defining a new trend that has emerged (see Graph 2). Does this change the odds? How do they compare with what the bookies are offering? Does a gambler have an edge on them because they have spotted something about the teams?

Regression analysis modelling on this level misses out a huge number of variables. Just because things are

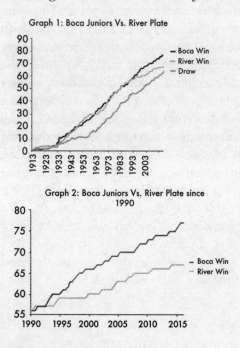

Graph 1: Boca Juniors Vs. River Plate

Graph 2: Boca Juniors Vs. River Plate since 1990

correlating, it doesn't mean there is causation between them. It looks as if River Plate may never win the derby again (but don't say that too loudly if you are in Argentina). This is where Big Data and more sophisticated models with greater predictive powers kick in. A company called BETEGY claims a 76 per cent success rate for global football results, going below the surface of mere goals scored, by taking factors including the weather and the birthday of the coach into account.

It's not only in the world of online football betting that sophisticated Big Data modelling is being used: Big Data analytics is going real-time. In the sphere of online poker, SharkScope, which claims to be the largest online poker tournament database in the world, is collecting vast amounts of data every day from millions of games so users can compare and track performance and avoid the sharks. Database queries that used to take twenty seconds to process now take only two seconds. Soon, you'll not only be able to choose your opponents, but also choose who to avoid. Even more daunting, an artificial intelligence poker program called Libratus pocketed $1.7m playing the no-limit game Texas Hold'em in a 20-day tournament in February 2017. It beat four of the world's top players in the process. How long it will be 'safe' to play online poker is now in doubt. Soon, we won't be able to tell if we're playing a human or an algorithm.

Big Data vs The Bookies

If the cuneiforms found in Mesopotamia are right, the domestication of pigeons began around 5,000 years ago. There is also a good chance the boast of, 'I bet my pigeon can fly faster than your pigeon' followed shortly afterwards. It seems we simply can't help ourselves when it comes gambling. All we've done is to move it from face-to-face human interaction to the anonymity of the Internet.

Where there's money combined with anonymity, there's also the possibility of criminality; there have been plenty of examples of sportspeople being tempted by betting rings. In 2010, snooker player Stephen Lee, once the world number five in the rankings, was banned from the sport for twelve years – effectively ending his career – for match fixing linked to suspicious betting patterns. The Pakistani cricket spot-fixing scandal of 2010 saw three members of the national team – Salman Butt, Mohammad Asif and Mohammad Amir – receiving bans from the sport after being convicted of taking bribes from a bookmaker, Mazhar Majeed, to deliberately under-perform during a Test match in London. In football, floodlight power failures in 1997 English league games, causing matches to be abandoned (rules state the score at the point of abandonment stands), were discovered to be the result of collusion between an Asian betting syndicate and security guards at the grounds.

Leaving aside the optimised gaming experience afforded by online gambling, it also has the added advantage of bringing formerly unseen and unregulated activity under surveillance and regulation. This has the further advantage of creating revenues more easily subject to taxation, a thought that most governments are more than willing to encourage. For this reason alone, online gambling has a sparkling future.

Whether you are a bookie, a punter or a service provider, Big Data is going to be part of the world of gambling in the twenty-first century. No matter how sophisticated the mathematics get, or how deep the models delve, it's likely that there will be room for the human eye to spot a break in the trend or pick up something a computer hasn't been able to detect. I wouldn't bet on it, though.

Never tell me the odds!

Big Data vs Money

Are we all turning into Larry Fink?

There is a good reason why economics is often referred to as 'the dismal science' – it is terrible at predicting things. Frankly, it is a wonder we have persisted with it for as long as we have.

Arguably, it's not the fault of economics per se. The famous economist Adam Smith introduced the idea of the market knowing best in his 1776 work, *The Wealth of Nations*. Smith believed there was a self-correcting mechanism within society, the 'invisible hand', restoring order if things got out of line. He articulated the ideas of supply and demand in a voice that is as familiar to us today as it was unfamiliar then:

> *When the quantity of any commodity which is brought to market falls short of the effectual demand, all those who are willing to pay ... cannot be supplied with the quantity which they want ... Some of them will be willing to give more. A competition will begin*

among them, and the market price will rise
… When the quantity brought to market
exceeds the effectual demand, it cannot be all
sold to those who are willing to pay the whole
value of the rent, wages a profit, which must
be paid in order to bring it thither … The
market price will sink…

The relationship between growth, inflation and employment was created within a simple set of 'truths'. The invisible hand would restore order if you just waited long enough – even if it was in the shape of a fist clunking down on the population.

Admittedly, it's a very appealing way of thinking. Smith's idea has resonated around the world for some three hundred years and is taught at all levels of education. However, you'll notice one thing – there are no numbers. No data. It's essentially a descriptive way of looking at the world rather than being scientific. Something that is scientific creates laws applicable in all space, at all times. Economics appears circumstantial, because the problems start for economics when you begin to look for the evidence in the form of numbers.

In the early days of economics, the lack of data could be forgiven. The nearest thing anyone really collected in terms of data was the census – the Romans' periodic survey of able-bodied men ready to trudge off to northern Britain along forgivingly straight roads. Over time, national governments started to collect data for the purposes of taxation, allowing economists to start constructing time series to see how things were going. Once we reached the point where data became reliably available, it became possible to test some cherished ideas such as the ones so confidently pronounced by Adam Smith. This is where things started to go wrong.

Take, for instance, the work of the New Zealand economist A.W. Phillips, who, as a student, looked at the relationship between employment and inflation in the British economy between 1861 and 1957. As Smith predicted, and Phillips found, inflation fell as unemployment rose and vice versa. This model came to be known as the 'Phillips curve'.

The problem was that there was no evidence the Phillips curve worked, or showed itself to be a permanently stable relationship for all economies, all over the world, at all times. This is no better illustrated than in the United States where, as the epitome of Smith's free-market ideas, it would be expected that the Phillips curve worked best. However, instead of a nice, smooth or even discernible descending line on a graph of unemployment versus inflation, something resembling a double-barrelled shotgun blast against a barn door revealed itself. Even economists turned on the Phillips curve. Debunking it turned into something of a cottage industry and the go-to subject if you

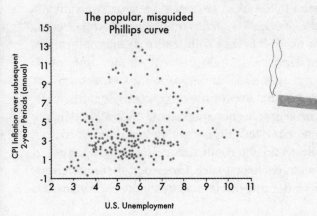

were aiming for the Nobel Prize in economics. Since 1974, seven Nobel laureates have been critical of some variation of the Phillips curve.

This is only one example of how economics and the real world don't match up. Despite the deficiencies, economics guides some of the most important institutions in society. It is used to set interest rates, manage social policies and direct the actions of fund managers, who place the savings of the great lumpen proletariat into the financial markets. So is it all dismal for economics? Can Big Data be used to illuminate the complexities of society to create a new kind of economics? What does Big Data mean to the people who manage our savings? What will be the future of how our money will be managed in the Big Data era?

Big Data and macroeconomics

Macroeconomics likes to make big pronouncements about the world. It studies indicators such as gross domestic product (what a country earns), unemployment, wages and prices, then looks at how they interact to understand how the whole economy functions. What macroeconomists like the most are models and forecasts. Given the slightest encouragement, they will slip some data going back tens, if not hundreds, of years, into a spreadsheet and set about predicting what is going to happen next using statistics. Or rather, that's what they would like to happen. In reality, macroeconomics has struggled to explain the world with any real consistency for nearly three hundred years.

The modelling techniques of macroeconomists haven't always been numerical – they have even extended to building physical models of how a society works. As a student at the London School of Economics, A.W. Phillips built the MONIAC (Monetary National Income Analogue Computer) to model the national economic processes of the UK. The MONIAC was an analogue computer using fluidic logic to mimic how an economy functioned. Fluid (meant to represent money) flowed through the system, pushing levers marked 'wages', 'taxes' or 'investment' up and down in a pleasing analogy of how an economy might work. Its weakness is that it saw the world in engineering terms: it is a deterministic system expressing the world as a series of cause and effect relationships with no randomness. As we all know, randomness is part of everyday life.

Here is where Big Data may have a huge effect on our understanding of how the world actually works, rather than how we would like it to work. It will require a shift away from interconnected levers and towards letting data speak through statistics and machine learning. Big Data also offers the possibility of taking information simultaneously from lots of variables to work out their combined effects on a single economic variable.

This shift is being helped by the appearance of large public and private data sets. For instance, a 2011 study using data from 2.5 million New York schoolchildren, over the twenty-five years after they left school, found the quality of their teachers had a profound effect on their future earnings. Replacing teachers in the bottom 5 per cent with an 'average' teacher raised lifetime earnings by a quarter of a million dollars. Meanwhile, a study using private data from 100 million eBay customers looked at the vexed question of Internet purchases and taxation. Does raising taxes in one area lead to consumers moving their purchases of the same item to a lower tax area, impacting jobs and the economy? Unsurprisingly, the answer was yes. Studies of this sort are able to lend credibility to arguments, through hard numbers and observed relationships, for the first time.

Researchers are also taking their work into real-time. Data is being collected about consumers during a browsing session, allowing retailers to perform experiments on small groups to test the effect of a new sales policy on consumer behaviour. The aim is to avoid some of the catastrophic mistakes that even the most legendary retailers have made in the past. In 1948, Maurice and Richard McDonald were gripped by the idea that they should change the formula of their reasonably successful San Bernardino drive-thru hamburger restaurant away from a customised set of almost limitless permutations brought to your car, to a limited menu which you collected from a window. Sales fell by 80 per cent. With some tweaks, business began to pick up again to create one of the most successful enterprises in the world: the McDonald's we all know today. It was a close-run thing, though. Imagine the sleepless nights the McDonald brothers could have avoided if they had lived in the Internet age and been able to use sales data gathered from their website, or had been able to experiment on a few customers to see the effect of changes before implementation.

Even the most familiar macroeconomic variables suffer from real-world problems. Take, for instance, inflation – the rate at which prices rise and fall. Inflation is watched keenly by the financial markets and policymakers. Too much inflation may mean an economy is overheating and needs to be cooled by raising interest rates. Too little inflation could be a signal that an economy needs a boost from lower interest rates. Lower interest rates help put money back into consumers' pockets through reduced mortgage costs and making it cheaper to borrow money – or rather that's how it's described in textbooks. In fact, inflation is a much more nuanced, observable effect and can be seen as a single variable with many different root causes: excessive amounts of money, currency devaluations, excessive growth, high wage growth and excessive money printing by central banks have all been blamed for high or hyperinflation rates. Economic historians can also usually come up with examples where all of these effects have been observable, but inflation hasn't been excessive. It really is a slippery eel.

This is where Big Data is beginning to help. The Billion Prices Project has been developed at the Massachusetts Institute of Technology (MIT). It relies on hundreds of websites over all over the world to calculate inflation in real time. Since online and offline

prices appear to be fairly close most of the time, researchers can confidently show that real-time data is a good predictor of the monthly data, sometimes released with a lag of several months, by various offices of national statistics around the world. It also offers the possibility to calculate inflation in economies where the data is either unavailable or unreliable.

The economy as a whole is merely a series of transactions between people with money and people with goods and services. A lot of the time those transactions are powered by credit, which makes the data within finance companies particularly interesting to macroeconomists. The financial services company Mastercard now offers a macroeconomic indicator service called SpendingPulse, driven by customer information which tracks transactions in areas as diverse as jewellery, electronic goods and motor fuel. Meanwhile, the payroll service provider Automatic Data Processing publishes a detailed breakdown of national and small-business employment trends, which is issued just a few days after the end of each month. The report covers 23 million employees in over 400,000 companies in the United States and is increasingly monitored by the financial markets as an important indicator in its own right, but also as a predictor of statistics issued by the US Department of Labor.

Today, most of the important economic statistics used by the markets are released every month, or every three months. Over time, we should expect that real-time indices of economic activity will become commonplace. Already, there are rich data sets to be found in Google Trends, while Twitter has been busy indexing hundreds of billions of tweets from 2006 onwards, covering human experiences and major historical events. It's easy to imagine that, one day, real-time economic statistics will replace the traditionally infrequent (monthly and quarterly) economic time series.

Big Data and the markets

In theory, economics and financial markets are deeply intertwined. It should work like this: when an economy is doing well, companies make higher profits, causing their stock prices to rise. It's reasonable to anticipate that interest rates will rise in the future, raising borrowing costs for companies and consumers, sending interest rates in the bond markets higher. At the same time, a booming economy, with high interest rates, could make the currency of a country attractive to outside investors – when demand increases, it causes the currency to appreciate against other currencies. Of course, the converse of all these benign effects are true: falling stock markets, falling interest rates and a declining currency should be witnessed at times of economic distress. What all these cause-and-effect relationships have in common is that they create price movements. Wherever there are price movements, there is the opportunity to buy and sell ahead of the crowd to make a profit. For some, it's just too tempting.

For this reason alone, the money markets are a natural home for data and analytics. Since the 1990s, the availability of financial market data has increased rapidly and it has become cheap to access. What used to cost tens of thousands of dollars a year can now be downloaded for free in real time (or with a slight delay). Yahoo! Finance and Quandl both offer the ability to download market data directly into spreadsheets for free. You could have a lot of fun analyzing price movements and maybe using some analytics to predict the next price movement – although it's not advisable to give up your day job to trade the market, as it's doubtful you'll make a consistent profit. There is simply too much information and too many variables acting upon the global financial markets simultaneously for any human being or simple trading system to handle it all at the same time. Enter Big Data and advanced analytics, in the form of high-frequency trading (HFT).

When it comes to trading the markets using rapid-fire data and advanced analytics, high-frequency trading beats everything else hands down. The use of computers and algorithms has now reached epidemic levels in the financial system. According to the Bank of England, HFT accounts for some 70 per cent of all trading on the US stock markets and 40 per cent of European markets each day, without any meaningful supervision. Fractional differences in time – measured in microseconds – dictate the difference between success and failure. HFTs fill the space between our ability to read a headline and any action we might take with hundreds of thousands of trades per second. It really isn't a fair fight any more.

The lengths to which HFT traders will go to gain a few microseconds of advantage is truly breathtaking (it takes eight microseconds to click a mouse button). HFT operators now have offices clustering around the art-deco façade of the carrier hotel (a large-scale data and communications centre) at 60 Hudson Street in New York. It houses the main telecoms arteries for the city, so getting physically close to it speeds up the rate at which you can trade. The New York Stock Exchange has gone so far as to issue standard-length cables to operators with connections to its trading systems to forestall accusations of unfairness. Even crude cables are starting to look obsolete, though. The new trend is towards traders operating at the nanosecond (a billionth of a second) level – the kind of speed it takes light to travel 29.98cm (11.8 inches). A series of laser beams and microwave transmitters now criss-cross Manhattan with one purpose only: to transmit financial trades faster.

Then there is the possibility of HFT causing a flash crash sometimes referred to as an ultra black swan event: unusual and rapid price movements which may last just a few seconds or minutes, wiping billions of dollars of value off companies, bonds or currencies for no fundamental reason besides a group of HFT algorithms feeding off each other's last trade. On the Albert Bridge which crosses the Thames in London, there is a sign telling soldiers they are not allowed to march in step, in case the synchronised thud of their boots sets off vibrational waves that could bring the whole structure down. It's a useful comparison. The system (like the bridge) can become unstable and chaotic when lots of small trades line up and reinforce each other. On 6 May 2010 at 2.45 p.m., the US stock markets experienced one of the most turbulent periods in financial history. It lasted for thirty-two minutes and the Dow Jones Industrial Average lost 9 per cent of its value, only to see it rebound and regain almost all of its value. In November 2016, Navinder Singh Sarao, an algorithmic HFT trader operating out of his bedroom in his parent's house in a west London suburb, pleaded guilty to twenty-two counts of fraud for his part in the US flash crash and for placing false trades which netted him an alleged $40m afterwards.

The controversy surrounding HFT focuses on three areas:

- What market purpose does HFT serve? Is it helping or hindering the process of establishing the 'right' price for a company?
- How does HFT generate high and consistent returns? What exactly is the anomaly that HFT is exploiting? Is HFT merely exploiting inefficiencies in the order system, rather than market mispricing?
- What social function is HFT performing? Is it right that HFT can cause excessive volatility in the basic commodities people rely on in their daily lives?

The effect of HFT and its dislocation of the markets from a fundamental economic reality has made dinosaurs of established money managers. It has even led some to throw in the towel. Well-known names in the fund management industry, such as BlueCrest Capital Management, Seneca Capital, and SAB Capital Management, have returned money to clients and added to an accelerating pace of funds shutting down globally. Big Data and the markets are having troubled beginnings, but once unleashed it's difficult to see how this particular genie could be squeezed back into the bottle.

Big Data and your finances

By 1986, Larry Fink had been working for ten years at First Boston, a large American investment bank. He had, by any objective measure, a spectacular career: from MBA graduate to managing director in a decade. Adding a billion dollars to First Boston's profits along the way. Then it went wrong: Fink and his bond department bet the next movement of US interest rates would go up in the second quarter of 1986. Instead, with the economy faltering and inflation falling, the Federal Reserve did the opposite and reduced interest rates. Fink and his department lost $100m. Suddenly, he was persona non grata and eventually left the firm.

Fink vowed to never let this happen again. He co-founded a company called BlackRock using a massive historical data set: the Asset, Liability and Debt and Derivative Investment Network, which neatly shortens to 'Aladdin'. It has 25 million lines of code monitoring over 2,000 risk factors and performs 180 million calculations each week. Fink built the largest fund management company in the world: BlackRock now manages around $5trillion on its own, and nearly 8 per cent of the world's total financial wealth helping others do the same.

Aladdin offers real-time risk assessments for a broad scope of investments, building complex portfolios using the results of thousands of scenarios, combined with modern portfolio theory (MPT). MPT was first introduced to the world by a twenty-seven-year-old economist called Harry Markowitz in 1952. His Nobel Prize-winning dissertation showed that if you used an enormous data set to look at the correlations between assets and events over time, you could create portfolios of investments which would survive in a wide range of circumstances and deliver to clients the holy grail of investment management: consistent and stable returns.

Besides offering the ability to create stable portfolios, the real genius of MPT is the ability of a client to choose what level of risk they feel comfortable with and then find the portfolio they should have. Or, to turn the problem upside down, define what kind of rewards they need and understand the kinds of risks they would be taking to get it.

For a long time, this sophisticated method of balancing risk and reward was only available to the largest clients, mainly institutions acting on behalf of pension funds and individuals of high net worth. Now what was once the haute couture of investing is available to all in the form of automated digital wealth management, sometimes called 'robo-advice'.

By asking a few questions about you, your lifestyle and your financial aspirations, robo-advisors can define your attitude to risk and the kind of returns you are going to need in the future. Then, using a Big Data set, it can design a portfolio and savings plan for you, which if all the correlations hold into the future, it expects will deliver what you want or need. Over time, as financial history unfolds, the correlations will change and evolve, requiring your portfolio to be rebalanced and tweaked. The idea is that you stick with your plan over the long-term, holding out the tantalizing possibility of bespoke tailored personal finances for the many. In August 2015, BlackRock and Larry Fink purchased FutureAdvisor, a digital wealth management company, to add to the side of its Aladdin risk management program. When a large financial company does that, you know that Big Data and sophisticated analytics are going to be part of all of our financial lives in the future.

Big Data vs Money

However you want to look at it, economics, the markets and personal finance are all engaged in a deep desire for greater certainty. The money game has become about avoiding mistakes rather than making bold decisions. We have all, in effect, become risk managers – smaller versions of Larry Fink.

This comes at an interesting moment in economic history and the changing relationship we have with money. The global financial crisis of 2008 wasn't only a shock to our financial systems: with some justification, some people saw it as signalling the end of the effectiveness of macroeconomic thinking and market-based mechanisms, as it failed to predict the crisis. During a briefing by academics at the London School of Economics in November 2008 about the turmoil of the international markets, Queen Elizabeth II asked, 'Why did nobody notice it?' It took four years for an economist at the Bank of England to answer her question.

The central bank remedies that followed the crisis – ultra-low interest rates, combined with the programme of market interventions called quantitative easing – saw massive amounts of money pumped into the global financial system and, in the process, buying-up financial assets at any price. Although the interventions unarguably saved the financial system from complete collapse, the casualty was the rational market mechanism, which is the bedrock of capitalism and modern economic thought. Without the possibility of loss sponsored by the world's central banks, the underpinning of asset prices (real estate, stocks and bonds) created what is known as 'moral hazard'. Moral hazard occurs when one party knows more than the other in a transaction and assumes more risk than they usually would. In other words, they know when something is a sure thing. It's like being told what colour is going to come up on a roulette wheel before it happens. The sure thing here is that if things go wrong again, central banks will step in to stop markets falling. This is the opposite of free markets, making them irrational and loaded with unseen risks.

Arguably, the decline of economic rationality, soothed by the rise of moral hazard, should come to an end one day. We will emerge, blinking, into the daylight of a new world. This is where Big Data can play a substantial role. It offers the tantalizing possibility of a new kind of economics, one using old ideas with a modern twist. It offers the possibility of new sensitivities, responsiveness and early-warning systems when things are going off the rails. It offers possibility of personalised economics and finances – one based on patterns of behaviour, preferences and needs. Big Data could help us avoid making the mistakes of the past – but it won't stop us from inventing new ones.

A new day of economics!

Big Data vs Work

Do you have what it takes to be an 'Awesome Nerd'?

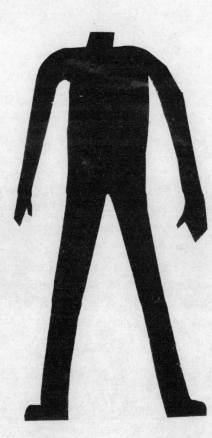

In early 1650, while on an ill-fated educational visit to Queen Christina of Sweden, the philosopher René Descartes died suddenly. In an act of supreme irony, his head became divorced from his torso when he was reburied in France. The two parts of Descartes were then sent to different places, thereby achieving the separation of mind and body that Descartes had so vociferously championed while alive and unified.

Descartes left the world after starting a huge argument with another philosopher, part-time lens grinder and instrument maker, Baruch Spinoza. Their spat revolved around a very simple dispute, which runs like this:

Me: I have brown curly hair.

Descartes: That's what you say. Let me have a look at you. Yes, I can confirm you have brown curly hair.

Me: I have brown curly hair.

Spinoza: I'm slightly busy right now looking at this kitten in a teacup on the Internet, so I'll accept you have brown curly hair for now, but I will check later.

[several minutes pass]

Spinoza: Ah! Yes! You do indeed have brown curly hair and isn't this kitten in a cup BRILLIANT!

There's an important lesson here. When do we make up our minds? Descartes said that we take information in, check it, then say whether we agree with it or not. Spinoza said that we accept the information as being true, then check it. It's not only philosophers who think this is an important distinction. Whether you favour the blind acceptance of Spinoza, or the immediate scepticism of Descartes, explanation has an effect on how we make decisions and how effective they are. The fact is, human perception, when faced with information and decision-making, is hopelessly flawed. We might act too quickly or not quickly enough. This is where Big Data can help. Is it even possible to remove the Descartes-Spinoza dispute by using Big Data and advanced analytics to reduce the uncertainties of decision-making in business?

How are businesses using Big Data to improve their performance? If you want to become one of the new breed of data scientists, what skills do you need to become one? How could this new business model affect work and society in the future?

Mission possible

Failure in business is part of the process of creating a dynamic capitalist system. The Museum of Failed Products in Michigan is the repository of glorious failure. Some on display include Clairol's 'Touch of Yogurt' shampoo, Heinz Purple Ketchup, ThirstyDog! and ThirstyCat! – beef and fish-flavoured fizzy water for dogs and cats respectively. If being in business has a mission, it is not be one of the 99 per cent of new products that fail and most importantly not to end up in the Museum of Failed Products. Using computers and Big Data can help that quest.

Using computers and data to manage businesses isn't new. Fred Smith's disarmingly simple name didn't stop him from writing an alarmingly forward-looking undergraduate paper at Yale in 1962, outlining an overnight delivery service in the computer age. Legend has it that he received only a C-grade for his efforts – but in the process, the paper led to Smith creating the courier company Federal Express (now known as FedEx). Sometimes, we don't need computers to test new products – we just need computers to identify harbingers of failure. An MIT study of 10 million transactions showed that there are some people with the unerring talent to pick certain items – ones that will definitely fail. What's more, the more they buy of a new product, the more likely it is to fall flat on its face.

Nearly all businesses gather data on their performance for monitoring purposes. Business intelligence (BI) exists to improve and optimise a business. Most companies treat data gathered for BI as something of a nuisance, as it's a tedious job to collect it and store it. Big Data aims to turn BI into a valuable commodity in itself. The volume, variety and velocity of Big Data available in modern businesses extends the analysis much further beyond static month-end BI and towards an analysis that shapes and changes a business, spotting new emerging trends and opportunities.

This requires the gathering and storage of enormous amounts of data. The happy coincidence of rapidly falling data storage costs and the maturity of Big Data analytics is now making this possible at an affordable price, bringing it within the grasp of even the most humble of companies.

The new business model uses Big Data going back many years in real time, or as near to real time as is useful. Transaction data (structured data) – who bought what, when and how – is combined with 'unstructured data' – comments from social media, customer feedback, sales meeting notes – to create a more complete picture of how the business is behaving. Business latency (the lag between action and response) is reduced, allowing companies to adjust product lines and prices to accommodate a shifting client base. BI, in this new manifestation, seeks to drive the business and maybe make it change direction by providing new insights into how consumer behaviour is changing and make those changes happen automatically – maybe without the intervention of human beings. Machines begin to run critical processes such as pricing and the Descartes-Spinoza debate simply disappears, along with a lot of jobs.

This opens up a whole new avenue: data and insights as a commodity, and a valuable one at that.

Once the culture of Big Data is embedded inside an organization, the insights gained can also be sold on to other tangentially related businesses. For instance, grocery stores will see much more timely behavioural changes than any economic indicator collected by national governments. Are people tightening their belts and buying minced beef rather than steak? Is the rise in luxury chocolate purchases outside of the usual holiday seasons a sign of confidence?

We really shouldn't forget the 'Internet of Things' – the interconnectivity of everyday appliances reporting back to base on how, when and the frequence of using a refrigerator, washing machine, toaster or television, among other objects. The collection and analysis of the generated data offers the opportunity for consumers to affect products and services. Again, companies can use or sell those insights to others. Using Big Data lowers the veil that stands between a company and its customers in a way never seen before. It allows an entry point into the kind of nirvana capitalism has long craved – business optimization and risk reduction in real time. It also offers the opportunity to give entrance to the Museum of Failed Products the body swerve.

The rise of the Awesome Nerd

In 2008, somewhere on the trading floor of the investment bank Lehman Brothers, someone pressed the wrong button. Or maybe it was the right button. Finally, a halt was called to the days of debt-fuelled consumerism: the financial crisis had begun. It nearly took with it the western banking system and the hopes, dreams and aspirations of a generation. Fortunately, the central banks of the world threw so much money at the problem that the dollar bills would have covered Manhattan enough times over to reach up to your knees.

It also precipitated another crisis. A lot of people had been hired to trawl through data using algorithms and databases of mortgages and loans so they could be sliced and diced into packages of debt that could then be sold on, in a process called securitization. The triple-A credit rating attached to them gave investors tremendous confidence and they sold by the bucket-load. The fees and rewards for the organizations creating the bond packages were astronomical. What's more, it seemed as if it could go on for ever.

In the process of nearly vaporizing the western financial system, they had given birth to industrial-scale data processing that attracted some of the best minds in the world. Those people had been plucked from science departments and engineering schools, dropped into offices and paid, in a year, what their professors could only dream of earning in a decade. When the crisis came, they were out of a job.

With almost freakish synchronization, the collapse of the securitization market coincided with the rise of what Harvard Business School called 'the sexiest job of the twenty-first century' and others called the 'awesome nerd': the data scientist. The job of the data scientist falls inside a Venn diagram intersection between mathematics, engineering, computer science and hacking. Demand for people with skills in the right areas of data science rose by up to 800 per cent in 2015 alone.

Thomas H. Davenport, in his excellent book *Big Data at Work: Dispelling the Myths, Uncovering the Opportunies*, provides some important traits you will probably need to display if you want to become a data science awesome nerd:

Code like a hacker: Awesome nerds need to be able to write computer code and understand how the technology of Big Data works (databases, Hadoop, Java, Python, SQL, NoSQL, Linux, etc.) with all the skills of those who seek to infiltrate and corrupt systems: hackers. Without the ability to code to a high level, you won't be able to put your ideas into action.

Think like a scientist: Scientists use evidence-based decision-making, but are also flexible enough to bend when the evidence changes (this is why lots of scientists are also musicians – they know their part, but can improvise around it). They also make excellent awesome nerds. Scientists tend to be somewhat impatient and want to do something with their findings. Action, rather than merely monitoring, is part of their make-up.

Be a trusted advisor: Awesome nerds need to be able to communicate with people more senior than themselves, but who may not understand the technicalities of what they are doing. Visualization of Big Data is also key: for some reason, we like to see Big Data in images. Don't forget, a company may re-engineer an entire process based on some awesome nerd finding, so you'd better understand decision-making processes.

Work like a quantitative analyst: Statistical analysis of data is a key part of being an awesome nerd. You will have to extract meaning out of structured data (transaction information) and unstructured data (video, text, images) using machine learning and artificial intelligence to design systems and processes to illuminate what is going on, but also spot business anomalies that could be a threat or an opportunity to the business.

Become a business expert: Even though you are an awesome nerd you can't sit in an ivory tower of your cleverness, illuminated only by the glow of your computer monitor. You need to be able to understand the business you are in and how it makes money. You also need to be able to work out where analytics can be used most effectively in the business.

If you are all of those things (or even a couple of them), then you are a rare beast indeed. The rewards can be handsome. In the US and Europe, average salaries for awesome nerds can be in the region of $110,000, with total packages, including bonuses and equity stakes, topping $170,000. If you get into management, this jumps to over $200,000 a year, while the upper end of director level hits nearly $300,000 per annum.

If this wasn't enough, as an awesome nerd your future appeal is also going to be broad, with your pool of potential employers growing by the day. They include:

- Any industry that moves things
- Any industry that sells things
- Any industry that employs machinery that can emit data
- Any industry that has facilities that can emit data
- Any industry that uses or sells video or musical content
- Any industry providing a personalised service
- Any industry that involves money transactions

So far, there have been some notable over and underachievers in the use of Big Data, AI and connectivity. Big Data teacher's pets include insurance companies, anything bought or sold online, credit-card transactions and consumer products in general. Meanwhile, the dunces in the corner of the Big Data classroom include banks, telecommunications firms, entertainment, shops and utilities.

Potential data scientists are being plucked from a range of disciplines – mainly the natural sciences – and they are more than likely educated to masters or doctoral level. The main reason for cannibalizing other specialties is that there are precious few degree courses specifically dedicated to data science at this time, although this is bound to change in the future. Some companies, such as Dell EMC, are developing internal programmes to grow their own data scientists to fill the gap that the lack of university courses has created.

Who will watch the watchmen?

Before you get to be an awesome nerd, you may have to get past the most formidable obstacle in your career: a company's human resources department (HR). The phrase 'human resources' has a paradoxical dehumanizing effect: treating human beings as a 'resource' should go against a job which is about dealing with people in all their individual complexity. Still, this hasn't stopped HR from embracing Big Data and predictive analytics with more enthusiasm than might appear seemly. Screening candidates for their performance before they have even joined a business has become both routine and controversial. Some 72 per cent of all CVs aren't even seen by human eyes any more. One step further are companies such as Gild, which sorts through the social data of millions of employees to pick out future star employees and even predict when they might leave. One of the biggest expenses associated with a workforce is the cost of replacing a worker: it usually adds up to 20 per cent of their base salary. Knowing the dynamics of your workforce could potentially save you money. Saberr is a company that will calculate your team's 'resonance score' to predict team performance.

This is controversial, because the process of 'human curation' can exclude specific groups from jobs without them ever knowing they are being rejected, besides a cursory email. Quantize, an early stage start-up company, applies machine learning to online CVs to identify the hard and soft skills to investigate applicants' deepest values and match them to existing top-performing employees in a company, completely side-stepping the data gathered when two people look at each other across a table. The resulting 'resonance score' determines whether you will enhance or detract from the team. Websites such as Klout.com use signals from more than 400 signals from eight different networks (Twitter, Instagram, Facebook, and so on) to calculate your 'Klout score': a measure of your influence. It has been used to screen employees for jobs. Someone with a Klout score of 37 may not even get past the door for interview. Former US President Obama has a Klout score of 99, while Justin Bieber is hot on his heels with 92. My Klout score is 10.

Systems are only as good as the data driving them. If an algorithm places a greater emphasis on the known and excludes the possibility of personal growth and change, it will exclude potential candidates. Just classifying yourself as being 'inquisitive' on a multiple-choice questionnaire could be enough to lose you the possibility of a position, because inquisitive people tend to seek out new questions and new opportunities. Some people are concerned that human curation and prescreening reverses the trend of blind testing developed to find undiscovered talent. Since the 1970s, orchestras, which have long been dominated by men, have routinely auditioned new players by asking them to play behind a screen – the music should speak for itself. Some people worry that the use of Big Data and predictive analytics in HR functions brings down that screen again.

From 'I, Robert' to 'I, Robot'

The rise of the awesome nerd isn't a new phenomenon: the world of work has never stopped changing and evolving as the result of technological innovation. The Luddites famously protested the 'fraudulent and deceitful' use of weaving machinery to get around standard working practices, literally smashing their way into the history. And witness the fate of the coachmen displaced by Robert Stephenson's Locomotion No. 1, the steam locomotive used for the first public railway journey from Stockton to Darlington in the north-east of England in 1825. The principles of business have never really changed: selling the largest amount of goods at the highest price you can get away with, while employing the smallest number of people necessary. What's left over is your profit.

For this reason, and with a few exceptions, business has always been about profit maximization through minimizing its largest directly controllable costs – employees. Salaries, wages, healthcare, pensions and places for them to work and sit usually make up about 70 per cent of the costs of a business. Minimizing employee costs by using technology is an inescapable obsession for business. We have always been trending from 'I, Robert' to 'I, Robot' – only the vehicle has changed.

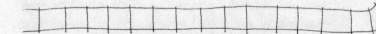

There are growing concerns that Big Data and its application through artificial intelligence and machine learning will have a profound and disruptive effect on both work and the distribution of wealth in the twenty-first century. The well-documented break in the distribution of profits from labour to the owners of capital, beginning around 2000, is now being intensified by Big Data and its applications. Workers who saw their share of economic output beginning to decline, while corporate profitability simultaneously rose, were only experiencing an acceleration of a trend which began in the 1970s: the income share of the top 0.01 per cent has risen from about 1 per cent to over 5 per cent in countries such as the United States. There, the top 0.01 per cent now earns more of the available income in a year than they did in the heyday of the 1920s, prior to the stock market crash of 1929.

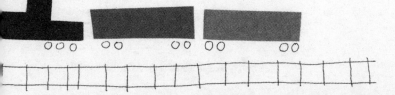

Rising income inequality is occurring even before you consider the effects of straightforward automation that is happening in even the most mundane and traditionally labour-intensive corners of life (see graphs overleaf). Take, for instance, Zume Pizza, based near Palo Alto in California. Zume uses robots to make pizza. Everything, from rolling the dough to spreading the tomato base and placing the pizza in the oven, is accomplished without the use of a single bored teenager. It's a remarkable achievement. In the background, there will be Big Data, collating customer transactions, feedback and experiences. So far, the vehicles the pizzas are delivered in are driven by humans: it's only a matter of time before Big Data-controlled self-driving delivery vans are carrying your Friday night treat.

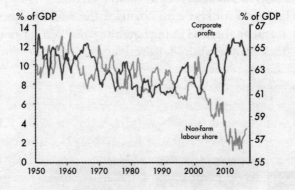

Non-Farm Labour and Corporate Profits Share of GDP 1950–2016

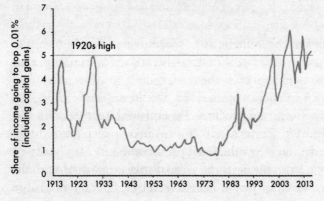

Top 0.01% Income Share 1913–2015

The development of automated vehicles (AV) will more than likely cause disruption to more than pizza delivery. Bus drivers, truck deliveries, tractor drivers, taxis, chauffeurs and self-employed drivers will all be impacted. The US Bureau of Labor Statistics estimated 3.7m jobs fall into these categories. These jobs also include critical driving and non-driving tasks, to varying degrees. Take away the driving element and what is left? Salespeople could spend time doing paperwork while a car drives itself, so AV wouldn't threaten those jobs directly. Taxi drivers and chauffeurs would be heavily impacted, though. A report published in the last days of the administration of President Obama concluded that between two and three million of the current driving-related jobs would be eliminated, or drastically altered, by the introduction of AVs. In February 2017, Ford announced it was investing $1bn in Argo AI, a start-up company dedicated to developing autonomous vehicle technology. General Motors acquired Cruise Automation, a company that makes specialised sensors enabling conventional vehicles to drive themselves on highways. Mercedes-Benz has already developed vehicles that can navigate unaided around European roads and in 2015 announced its F 015 concept vehicle – a 'moving living space', which is its blueprint for the future of driving. What was once science fiction is now becoming science fact.

Of course, other jobs would be created in the process, in areas such as supervision, maintenance and customer services. However, anyone displaced rarely, if ever, recovers the level of income lost. Ten years or more after losing their jobs, displaced workers usually have earnings that are depressed by 10 per cent or more relative, to their previous wages. Moreover, and critical to the government finances which are entirely dictated by the revenue they take and the money they can borrow, robots don't pay taxes.

Big Data and AI is going to bring many benefits, but the transition to a more data-driven, evidence-based society is going to involve considerable pain for some people, especially in the lower income bracket.

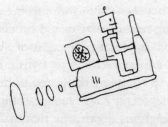

Big Data vs Work

Denmark is, officially, the happiest place in the world. They even have a word for it: *hygge* – a kind of contentment of the soul – and institutes such as the Happiness Research Institute in Copenhagen are dedicated to defining what it means to be happy. More to the point, Danes seem to understand that the reason for their happiness is the balance between life, work and the role of the state in their lives. Eye-watering levels of taxation are tolerated, because there is a social contract that essentially says that the state will catch you if you fall. Combined with an enviable work-life balance, Danes are also the calmest people in Europe. According to the European Social Survey, 33 per cent of Danes report feeling calm and peaceful most of the time. This compares to 23 per cent in Germany, 15 per cent in France and just 14 per cent in the UK.

Big Data and automation, in theory, offer the opportunity to tend towards Danish levels of contentment. There has been a global downward trend in the average hours worked since the 1950s (the US appears to have stalled in this respect in the early 1980s), providing the opportunity for improving the work-life balance for many people.

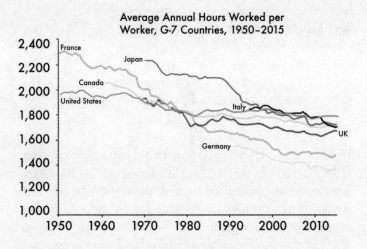

Average Annual Hours Worked per Worker, G-7 Countries, 1950–2015

What hasn't yet happened is a fairer distribution of profits from productivity gains: so far, it has only resulted in a generalised decline in wages, as more people compete for fewer jobs. The imposition of the minimum wage has gone some way in attempting to redress the balance of income inequality, but it has only so far been sufficient to slow the rate of decline in relative living standards, rather than restore it back to where we were as recently as the 1970s. Other ideas, such as the Universal Basic Income (UBI), which is a payment to every adult and child irrespective of circumstance is, so far, too advanced for many politicians to contemplate – although there are advocates of it from both the left and right of the political spectrum and surveys show it is a popular idea among voters. Finland and Alaska are places where a form of UBI has been trialled, but examples are few and far between (except as pilots), even though the forced redistribution of wealth makes economic sense, as it raises consumption and the gross domestic product (income) of a nation.

Most economists agree there will have to be a response to Big Data – and there will be a role for governments to play in three key areas:

Invest and develop for the benefits of Big Data and AI: Accept and acknowledge the Big Data workplace is coming and invest through research and development grants to encourage responsible and ethical AI technologies.

Educate and train for the future: The Big Data workplace changes the educational and training needs of the workforce. This starts with providing all children with access to high-quality education to prepare them for a lifetime of learning. Retraining may become as much a part of life as changing jobs.

Help workers in the transition and provide a safety net: Strengthening critical support structures, such as unemployment insurance, medical care and emergency aid for families when AI-driven job displacements occur.

There might not be anything like a Dane, but that doesn't mean we can't all aspire to it in the new Big Data economy.

Big Data vs Politics

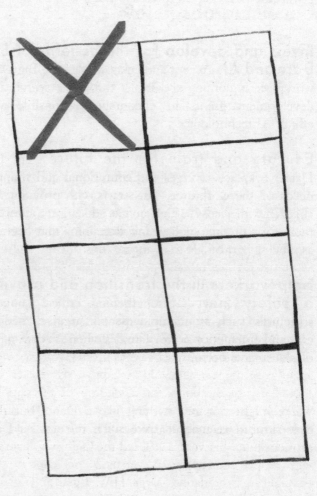

Are politicians becoming obsolete?

During his 1992 bid for the US presidency, Ross Perot cut an unusual figure. The sixty-two-year-old ex-US Navy officer had all the gravitas of a generation of people who you would have expected to favour dealings in the smoked-filled backrooms of private clubs in Washington DC, rather than offering strong advocacy of a new wave of democracy.

But the former IBM salesman, who had created a billion-dollar computer business of his own, was an enthusiastic supporter of a real-time 'electronic town hall' as a solution to the country's problems. Perot went as far as to suggest he would be happy to be sacked at a moment's notice, should the fickle flow of the people's will turn against him.

The idea seemed to have some resonance at the time. In June that year, Perot led the Gallup poll survey with 39 per cent approval (versus 31 per cent for the Republican candidate George H.W. Bush and 25 per cent for his Democratic rival, Bill Clinton), a result made all the more remarkable by the fact he was standing as an independent candidate. He eventually slumped to 19 per cent of the popular vote, winning no electoral college votes, mainly due to a chaotic campaign during which he was, ironically, accused of not being willing to listen to advice.

Perot's vision of a digital democracy envisages a never-ending round of referenda, open to all. It relies upon the existence of an enlightened and intelligent electorate, willing to come home from work and do the necessary reading to make informed decisions, instead of maybe falling asleep in front of the TV. Karl Marx would no doubt have applauded the aim of reconnecting the alienated social classes back with their own destiny, even though it was surrounded by the whiff of Perot's main aim, which was to sell more computers to the general public.

His vision was far-sighted by any measure. The Internet was still in its infancy and the release of the first version of the revolutionary Netscape web browser was still two years away. Mobile phones still had extendable antennas. It also presumed a fully engaged electorate, when all the evidence pointed in the opposite direction. Political party membership and turnout out at elections had been falling consistently since the 1970s in the US and Europe. Even Russia, after an initial enthusiasm for voting following the dissolution of the Soviet Union in 1991, has seen its participation rate slump from 80 to just 40 per cent of the voting age population. At the current rate, it won't be long before it descends below the 40 per cent rate of the European Union as a whole. In the 2016 US presidential election, 43 per cent of eligible voters didn't vote.

Given the advances and spread of technology since 1992, is there a new digital democracy on its way? What does Big Data mean for politics in the remainder of the twenty-first century?

Opinion poll dancing

Opinion polls are more noticeable for their failures than their successes. The *Chicago Tribune*'s premature prediction, based on a Gallup opinion poll, that Thomas Dewey had beaten Harry S. Truman in the 1948 presidential election, was only the beginning of a long series of humbling experiences for pollsters at the hands of the general public. Brexit, the UK's exit from the European Union, was missed until the votes were counted. Donald Trump's victory in the 2016 presidential race was predicted by only a few pollsters, most of them using non-polling techniques. Significantly, the two polling groups who correctly predicted the Brexit result consistently conducted their polls via anonymous online surveys. It precipitated the latest crisis in methods used by pollsters and made some wonder whether they would ever recover their credibility.

Traditionally, opinion polls are conducted by a simple phone call or online questions. Organizations such as Gallup even go as far as to conduct a daily 'world poll', asking 1,000 citizens in 160 countries a series of 100 questions about law and order, food and shelter, jobs and their wellbeing.

The problem that opinion polls encounter include human frailties, such as being susceptible to the way questions are posed, how the question is phrased and the sample of people chosen. Careful choices of the interrogators and sentence construction ('*Would* you vote for candidate X?' is a somewhat different question to 'Would *you* vote for candidate X?', or 'Would you *vote* for candidate X?').

Despite the sampling problem, opinion polls are usually reported in the press as a definite number, but actually come with an accuracy warning. It's not unusual for polls to suffer a margin of error of plus or minus 3 to 4 per cent. A quoted 48 per cent for a candidate could actually be +52 per cent (48 + 4), or +44 per cent (48 − 4). What's more, polls are calculated with what is called a 95 per cent confidence level. If the poll was performed 100 times, it would come in within the margin of error, 95 per cent of the time. This isn't the same as saying it is 95 per cent accurate: it's actually saying that, 5 per cent of the time (one time in twenty) the polls are going to be wrong by something larger than the error. Given that most political polls in two-party systems are won or lost on the opinions of about 5 per cent of the population, it is frankly a miracle that opinion polls even come close to predicting the correct result as often as they do.

To add to the woes of pollsters, some worry that polls are becoming increasingly inaccurate because of technology and people wanting to say what is socially acceptable, rather than what they actually think. The effects of technology were cited in the surprise victory of the Likud party of the embattled Israeli Prime Minister Benjamin Netanyahu in 2015. The use of mobile phones prevented pollsters tracking down the randomly chosen respondents, making their carefully constructed sampling methods almost meaningless leading to a shock result. This is never more the case when call screening means you don't have to interrupt your latest round of Candy Crush Saga. Both engagement and participation is now a problem for pollsters.

People have stopped telling the truth. There is a growing realization that social conditioning, the process of training individuals to respond in a certain 'approved' manner, has been circumvented and replaced by social desirability bias – the tendency to tell surveys what they want to hear and then do something else in the privacy of the voting booth. Some think the effect is beginning to render traditional polling methods redundant. Something more sophisticated may be needed to tease out the true feelings of the electorate in the future. Enter Big Data.

I tawt I taw a puddy tat a-kweeping up on me...

As far as traditional polls are concerned, it's not so much Sylvester the Cat creeping up on Tweety Pie as much as it is tweets and social media stalking the pollsters. Traditional polling relies on comparing the 'this time around' with the 'last time around'. It can't take account of rapid changes or great leaps in opinion, whereas Big Data, coupled with sophisticated analytics, revels in these discontinuities.

As early as 2010, researchers at Carnegie Mellon University had been scraping data out of Twitter using keywords to test whether they could recreate polling data taken from economic and political surveys. Even allowing for timing differences (they were essentially working in real time, but surveys are reported days or weeks after the data is collected) they were able to reproduce and predict US President Obama's approval ratings with an 80 per cent accuracy. Allowing for a doubling in the use of Twitter over the two years of the study between 2008 and 2009, during which the researchers collected a billion messages at a rate of between 100,000 and 7 million a day, it was a remarkable result.

On this evidence, the old days of waiting for the reinforcement of approval at the ballot box every four or five years may soon be a thing of the past for politicians. Companies such as BuzzMetrics offer real-time monitoring of consumer brands, harvesting data from local and global blogs, news sites and social media. Using natural language-processing, marketing companies are now not only monitoring trends, but also seek to identify when a potential social media crisis is brewing long before it happens. Responses to shifting public opinion, public relations and politics are going real time.

Big Data and the end of politics

Shaping social policy using data, effectively taking it out of the hands of politicians, isn't a new idea. Charles Booth exemplifies how analysis has been used to shape society for a long time. Booth had become disillusioned with politics, having been rejected by the electorate at the 1865 British general election. Instead, he realised he could do more good working outside the system, using a Victorian equivalent of Big Data as his weapon. Booth set about creating a detailed map of London which would highlight specific areas of poverty among the population. It was published as part of his multi-volume study, 'Life and Labour of the People in London', with the intention of reforming the living conditions of Londoners.

Booth and his researchers tramped the streets of the city, categorizing households by income, to produce a colour-coded map of poverty – what we might call an infographic today. What Booth found was shocking, especially if you fell into the black-labelled 'Vicious, semi-criminal' category. Along the way, Booth managed to coin the term 'the poverty line'. Eventually, slums were cleared and a new era of social housing was beckoned in. It was a remarkable achievement for someone who had made his fortune from making gloves.

What Booth was doing in the nineteenth century, we are doing today. Instead of walking the streets of a city, we now use the mass collection of electronically sourced data to micro-manage society. It took Booth and his assistants, Clara Collet and Beatrice Potter, four years to create their map. We are now attempting similar projects in real time.

This is partly a response to humans becoming increasingly urbanised. Cities are the place of choice to live. Research by the University of Washington suggests that, 'In the thirty years since 1979, China's urban population grew by about 440m to 622m in 2009'. It's a trend that is being echoed around the world. Using Big Data, some cities have found it possible to form policy and respond faster than the electoral cycle. In some places, notably in the US, it is being positively encouraged. In the city of Boston, drivers are now able to download an app that registers if their car hits a bump in the road. By feeding this back to a data centre, the authorities are able to build up a real-time heat map of road problems in the city and respond in a timely fashion. Los Angeles is replacing 4,500 miles of streetlights with smart LEDs controlled by Big Data. The new brighter lights form an interconnected system, informing the city of each bulb's status. Broken bulbs can be identified and fixed almost immediately. In the French city of Lyon, sensors monitor traffic congestion to allow the free passage of emergency vehicles. By 2018, New York will have over 200 data sets released on a daily basis, covering anything from traffic violations to building permits and school attendance. The New York Police Department plans to release annual data showing precinct-based comparisons of 'stop and suspect' descriptions, an annual report on firearm discharges and weekly city-wide crime statistics by borough. The sole aim of this initiative is to provide data scientists with free data that can be read and analyzed, sometimes in real time, to come up with new solutions outside of the government process. Politically, this takes away an opportunity for politicians. The roads in Boston will no longer get to a cataclysmic point, so 'Vote for me to fix the roads!' disappears from the electoral stump.

The message for politicians is becoming increasingly clear: using Big Data and algorithmic decision-making to track and respond to the lives of citizens sets up a feedback system, turning cities into responsive organisms, flexing and changing with their occupants' needs. The value of using social change as a bargaining chip in a political process becomes significantly reduced. Some social analysts see this effect as the beginning of the end of traditional, ideology-driven politics, as it cannot survive the hot, probing eye of sophisticated, evidence-based analysis. The phrase 'Trust me, I'm a politician' has never looked so out of date as it does in the Big Data era.

The fightback is already starting. A new trend in political campaigning is emerging, one which combines psychometric testing with the techniques developed in targeted advertising. Have you ever wondered what happened to all the data you generate finding out 'Which Game Of Thrones character are you?' and all the 'likes' you offer up on social media? Combining this data with credit card, transactional information and residential location, among other things, allows political researchers to guess the colour of your skin, your gender and your voting intentions. As they know where you live, you can now be targeted with individual messages, tailored to make the campaigning funds go further. The age of blanket mass communication may well be over. The targeted communications created by the company Cambridge Analytica, using Big Data and algorithms, have been credited with having affected both the Brexit vote and the election of President Trump in 2016. In the future, politicians will look more like the advertising bar on a search engine than the ideologues of the past.

You bought this, so you might like to vote for me!

Man vs Politics

After the fall of Athens, democracy lay dormant for 2,000 years until its re-emergence during the Enlightenment. A spirited defence against democracy by monarchists in the nineteenth century couldn't stop its rise. Even so, by 1941 there were only eleven democracies left after a number of spluttering starts. Since the end of the Second World War, democracy has become, according to the Freedom House think-tank, the dominant political model in 120 countries, or 63 per cent of the world in total. It's one of the twenty-first century's greatest success stories. However, some have begun to worry that we may have reached the high-water mark for democracy. Outside the West, democracy seems to advance only to collapse later on, while western democracies have become debt-laden and dysfunctional due to their over-reliance on making short-term promises and failing to plan for the long-term.

Coincidentally, China's centralised decision-making model has shattered the democratic world's claim as the only way to create economic prosperity. Chinese living standards are rising at a rate of multiples over mature democratic nations, while its willingness to recruit new talent and refresh its political establishment every decade is in stark contrast to the sometimes dynastic tendencies of the West.

At the same time, twenty-first century democracies have come under a double attack from globalization and localization. Globalization robs nation states of their autonomy, while localization takes away the power of national governments within their own borders. As a national democrat, you wouldn't look out of place on the endangered species list these days.

The response of politicians who are no longer able to dictate events is to content themselves with managing perceptions. Truth is the immediate casualty and word games replace substance, creating what has become known as 'the post-truth era'. Power becomes an end in itself. When the writer Umberto Eco said, 'There is in our future a TV or Internet populism, in which the emotional response of a selected group of citizens can be presented and accepted as the Voice of People', he was, in his polite Italian philosophical way, presenting a vision of the future not based on 'analysis', but 'feelings'. It is an almost nightmarish apparition, where perception management is much more important than evidence and truth. The effect can be amplified if individuals can exist in a resonating echo chamber of self-reinforcement. In pop culture, this is sometimes called 'death by yes' – the downward physical or creative spiral of artists who gradually rid themselves of the people who say 'no' to their more self-destructive urges. This is where Big Data may yet help to promote and strengthen democracy and the political process, by correctly gauging the authentic feelings of citizens and allocating resources where they are needed most. It may become the accepted method of running a society in the Big Data era.

The phrase,

'Trust me, I'm a politician.'

Has never looked so out of date

as it does in the Big Data era.

Big Data vs Science

As a species, is it worth getting out of bed any more?

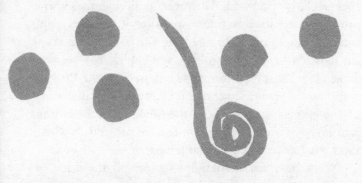

One of the few things we know for certain about humans is that we aren't scientists by choice. Being a scientist consists of starting off with an idea about how the world may work, designing an experiment to test it, then letting the results do the talking. During the process, it's best to build in a test to prove you could be wrong, just to be on the safe side.

This is not how humans instinctively investigate the world. Our default setting is to start off with a set of self-generated prejudices and then look for the evidence to reinforce that view. Anything that gets in the way is merely an inconvenient distraction from the main purpose of confirmation. This is why science is such a shockingly large leap of human imagination; it goes against all our instincts. Until the eighteenth century and the Enlightenment, this was our main source of sorting out the world. It led to a number of notable mistakes: an awful lot of 'witches' were drowned before we embraced the method of 'thesis, antithesis, synthesis'. In human terms, things really started to motor along after this particular light was illuminated in our heads.

As a way of going about things, science has been spectacularly successful. However, as all scientists know, proposing an idea is fine – proving it requires data, sometimes lots of it. Arguably, in the past fifteen years we have had our own golden age, which has seen us make giant leaps forwards: the Human Genome Project, mapping the genetic code of humanity, and the discovery of the Higgs boson particle (the final link in the atomic puzzle) have both been data-intensive projects in which Big Data has played a substantial part.

What now for mankind? Are we on the edge of personalised medicine because of genome-mapping through Big Data? Have we reached the end discovery in physics? Are these Big Data projects going to put an impossible strain on the global computing system?

'It's life Jim, but not as we know it'

It's almost impossible to imagine a time when we didn't understand how our personal characteristics were passed from generation to generation. Charles Darwin, in his seminal 1859 work *On the Origin of Species*, had at least introduced the idea that some form of inheritance of population variations helped the fittest to survive in the evolutionary struggle. The 'how' of the inheritance process was unclear, until Gregor Mendel and his peas came along.

Between 1856 and 1863, Mendel studied 8,000 edible pea plants, tracking how characteristics were passed from one generation to the next and concluding that there were discrete inherited units that gave rise to observable physical characteristics. In other words, genes. He was onto the idea that packets of information were being passed from one generation to the next. In 1903, two people on opposite sides of the Atlantic were simultaneously doing things to grasshoppers and sea urchins. Walter Sutton and Theodor Boveri, respectively, worked out that inheritance was being passed on via chromosomes in the nuclei of cells. In the 1940s and 50s, chromosomal deoxyribonucleic acid (DNA) was shown to be the vehicle in which genes travelled. It was James Watson and Francis Crick of Cambridge University who threw caution to the wind, interpreting the X-ray crystallography results of Rosalind Franklin and Maurice Wilkins and declaring the double-helix structure of DNA in 1953. The breakthrough was not just the double-helix shape (a kind of twisted ladder), but also the way the four identified basic units (cytosine, guanine, thymine and adenine) combined together in 'base pairs'. Cytosine always bound to guanine, and thymine always bound to adenine. It was the discovery that the combinations were so unalterable that was so shocking.

There are about 20,000 genes in a human being, each one made of arrays of base-pair letters such as CGGTCATTCATACG. This is the rule book which tells an eye cell to be an eye cell and a skin cell to be a skin cell. If you thought like a data scientist, you'd realise the three billion base pairs on two sets of twenty-three chromosomes in a single human cell means the human genome contains about 1.5 gigabytes of data. You could fit it on to two CDs.

The stage was now set to study human genes. The Human Genome Project (HGP) set out to sequence the three billion base pairs. On the face of it, the HGP worked in a pretty simple, if laborious, way. First, the genetic material was blasted into 20,000 small fragments (about 150,000 to 200,000 base pairs long). Then, using bacteria, each of the fragments was copied millions of times to create a kind of dense 'soup'. For the sequencing process, the fragments were cut into pieces about 500 to 800 base pairs long. A fluorescing version of the C, G, A and T was added to bind to their duller counterparts. Using a laser, they were excited to light up as they passed through a detector and a picture was taken. This identified the base-pair sequences on each fragment. This was done ten times. Then, using a sorting algorithm, there was analysis of where the sequences overlapped, so it could be stuck back together again to create a road map of the gene. This is known as the 'reference sequence'. It was an almost archetypical Big Data project: lots of data and an algorithm to make sense of it. You just had to do it a lot of times and have about twenty laboratories around the world working flat out for about thirteen years at an estimated all-in cost of $3bn. The HGP was declared complete in April 2003, with 92 per cent of the possible genome mapping verified to a 99.99 per cent accuracy.

The HGP used an unknown number of anonymous donors to create the reference set of genes. The first person to have his personal genome published was scientist Craig Venter. Venter had set off a race with the HGP because he and his company, Celera Genomics, thought they could complete the project faster than the laborious methods used by the HGP. He was right: the resulting collaboration shaved years off the time taken to map the genome. Venter was voted one of the most influential people in the world in 2008 by *Time* magazine, which must have been satisfying. Probably less satisfying was the discovery he had a genetic propensity towards antisocial behaviour, Alzheimer's disease, cardiovascular disease and wet earwax. The former of these was confirmed when Venter attempted to patent specific human genes – the ensuing outcry quickly saw him drop the attempt.

Since then, former archbishop Desmond Tutu and an entire family in Spain have also had their genomes sequenced. By 2011, we passed the 1,000 mark for individually sequenced genomes. As the price of sequencing declines, the number of completed genome sequences is doubling every seven years. There are now 2,500 high through-put genomic sequencers clicking away in 1,000 centres worldwide. This will put a huge strain on data storage in the future. Comparing the storage needs of Twitter, astronomy, YouTube and genomics out to 2025, researchers have shown that genomics may need up to 40,000 petabytes of additional storage per year whereas the other groups will need a maximum of just 2,000 petabytes each.

YouTube: 1,000–2,000 PB/year

Astronomy: 1,000 PB/year

Twitter: 1–17 PB/year

Genomics: 2,000–40,000 PB/year

As analysis techniques improve, genetic decoding will approach real-time sequencing, holding out the possibility of personalised medication. Joshua Osborn had baffled doctors for weeks with an inflammation that had dangerously swollen his brain. In a last throw of the dice, his doctors asked a team at the University of California to index fluid taken from his spine. Within forty-eight hours, they had their answer: Joshua had been infected by the bacteria Leptospira, which could be treated with a massive dose of penicillin. Two weeks later, Joshua was walking again. That was in 2014 and the diagnosis may soon look like it was achieved at a snail's pace compared to modern techniques. Already, programs such as the Scalable Nucleotide Alignment Program (SNAP) developed at the UC Berkley AMPLab compare a patient's genetic sequence to the reference genome to find potential problem-causing mismatches that is between three and twenty times faster than traditional computing methods without sacrificing accuracy. SNAP gets its speed from being able to read longer lengths of DNA fragments in the most likely locations, combined with increased computing power. The team hopes that cancer patients will one day routinely have their genome sequenced to find the most effective treatment for them.

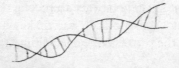

The real game-changer in personal genetics in the future is the CRISPR/Cas9 gene-editing process. CRISPR (clustered regularly interspaced short palindromic repeats) is like giving a three-dimensional Pac-Man

(Cas9) its own set of bespoke dentures. When locked exactly onto a matching length of DNA, it can then munch through it. The gene can then rejoin, containing a mistake (effectively switching the gene off) or the Pac-Man can insert a new, undamaged section of the gene it is carrying with it. This is engineering at the atomic level and it is, frankly, a miracle.

CRISPR offers a future where specific genetic diseases such as Huntington's disease, muscular dystrophy or sickle-cell disease could be wiped out. It also offers the possibility of increasing our knowledge of how the genome works and understanding what the 98 per cent of the human genome, which appears redundant, actually does. However, because it can be used to target and edit out diseases before conception, it also raises ethical problems. Could it eventually increase inequality through eugenics and discrimination? From a Big Data perspective, the massive processing power and storage required to embark on this route will only increase the load on our computing and storage systems in the future. The 40,000 petabytes of additional storage predicted for genome sequencing could be a drop in the data ocean should CRISPR/Cas9 technology become as commonplace as many predict. Its usability as a technique is only being enhanced by machine learning, which is now being used to reduce the accidental or off-target cuts the Pac-Man molecule can make, saving time and speeding up the process of gene editing by up to 50 per cent.

To get to this point, the journey had been a kind of evolution in itself. It may only be a matter of time before we have home genome sequencing via an app. It may save a lot of time and heartache if, on your first date, you could ask for a quick saliva sample to test for the AVPR1A – the so-called 'infidelity gene'.

Since we now know what we are looking for, both the time taken and the cost of gene sequencing is reducing quickly. When the HGP kicked off in 1988, it cost around $50 to sequence one base pair. That cost has fallen over time, but still it cost around $750m in laboratory time to create the first genome. Today, because the techniques and technologies have been perfected and refined, the cost has fallen to about $1,000. If you only sequence the exome (the 1 per cent of the genome where 85 per cent of all mutations occur that affect disease), the cost has fallen to considerably less than $1,000. Less detailed versions are also available. The process is as simple as spitting into a tube, sealing it and sending it to a company such as 23andMe, who, for about $180, will give you some interesting results. Here is my data:

- Ancestral composition (I'm 99.8 per cent European)
- How much of a Neanderthal you are (I'm 3.1 per cent, which puts me in the ninety-eighth per centile. 23andMe offers the opportunity to buy a '3.1% Neanderthal and Proud' T-shirt)
- Genetic risk factors (the 'bad' Alzheimer's APOE-ε4 variant, Parkinson's disease, breast and ovarian cancer variants were absent from my results)
- Inherited conditions (I'm lactose intolerant, which was news to me, but fits with my deep aversion to cheese)
- Drug response (I have rapid metabolizing of proton-pump inhibitors, so if taking drugs for stomach antacid, I may need more of it)

I am so British and Irish (70.8 per cent) that my genes have been walking around here since the Vikings, Saxons and Celts swept through. It looks as if the Romans didn't even glance at my village of Billingham, in the north-east of England, as there are no Italian genes that are detectable. On the face of it, I should be wearing a Union Jack waistcoat and weeping uncontrollably at the sound of 'Land of Hope and Glory'. Some of the data from 23andMe you can get by looking in the mirror (confirmed that I have blue eyes and brown curly hair) and some of it confirms your own suspicions (I am genetically a sprinter, not a marathon runner, with a slight preference for sweet foods). Some of it is just odd (I possess a slightly higher risk of 'photo-sneeze effect', the tendency to sneeze when moving from relative darkness into bright light). Ancestrally, I am offered the connection that I am of the same family line as the Russian writer Leo Tolstoy, author of *War and Peace*.

Stewart Cowley's Gene Map

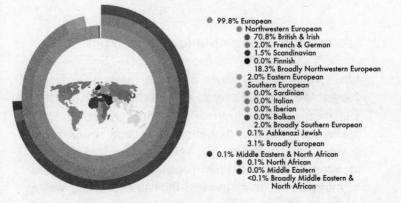

99.8% European
Northwestern European
70.8% British & Irish
2.0% French & German
1.5% Scandinavian
0.0% Finnish
18.3% Broadly Northwestern European
2.0% Eastern European
Southern European
0.0% Sardinian
0.0% Italian
0.0% Iberian
0.0% Balkan
2.0% Broadly Southern European
0.1% Ashkenazi Jewish

3.1% Broadly European

0.1% Middle Eastern & North African
0.1% North African
0.0% Middle Eastern
<0.1% Broadly Middle Eastern & North African

There is one final Big Data consequence of cracking the DNA code and our newfound ability to manipulate base pairs along the way: we could store data in cells. If you think of the two base pairs (C-G, T-A) as zeros and ones, a strand of DNA could act as a digital storage device if you converted the data in a digital file into a string of base pairs. To read the data, the DNA is sequenced, reversing the encoding process. It's an idea offering the possibility of huge amounts of data storage in small spaces with incredible stability – DNA can survive for hundreds of thousands of years. It could be the solution to our long-term data storage problems. In January 2013, the journal *Nature*, reported a group at the European Bioinformatics Institute had encoded over five million bits of information in a DNA sample the size of a speck of dust. It contained among other things 154 of Shakespeare's sonnets, a twenty-six-second audio clip of Martin Luther King's 'I have a dream' speech, and the original paper on the structure of DNA by James Watson and Francis Crick. The researchers were able to retrieve the files with between 99.99 and 100 per cent accuracy. The age of the biological computer is on our doorstep.

'Ye cannae change the laws of physics!'

On 4 July 2012, a rather uncertain looking Professor Joe Incandela stood up in an auditorium at CERN (the European Organisation for Nuclear Research) and announced, 'If we combine the z-z with the gamma-gamma this is what we get. They line up extremely well at about 125 GeV. They combine with a combined significance of five standard deviations'. There was a gasp, everybody clapped and the eighty-two-year-old Peter Higgs, Emeritus Professor of Theoretical Physics at Edinburgh University, wiped a tear from his eye. It was left to Rolf-Dieter Heuer, then the director general of CERN, to put it into normal words: 'As a layman I think I would say, I think we have it.' The Higgs boson's existence had been confirmed with a near certainty that only the most cautious of physicists would qualify. With it, the Standard Model of the atom was made real and whole. Less happy was Professor Stephen Hawking, who had bet that it didn't exist. He lost $100.

The Higgs boson matters a lot. Much was known about the Standard Model of the atom: there are twelve fundamental particles called fermions and four force-carrying particles called gauge bosons and many of the interactions were understood. A nagging mystery remained. What gave the particles mass (weight)? This is where the Higgs boson came in. In 1964, Peter Higgs and six other scientists proposed that subatomic particle mass came about as a result of them passing through an energy field. It was the interaction with the field that conferred the mass on the particles. Think of a fish swimming through water: the bigger the fish, the more interaction (resistance) it has with the water as it swims through. Particles interacting strongly with the field would have a large mass, while things that essentially passed straight through would have no mass.

If the field existed, what was it made of? Just as water is made of water molecules, Higgs and his colleagues proposed the Higgs field was made of particles (Higgs bosons), which would have a very particular mass itself. If you could find a particle with that mass, you had found the Higgs boson, the thing joining everything together. Hence, it was dubbed the 'God Particle'.

To find the Higgs boson would require breaking atoms down to their smallest possible units and somehow record if a Higgs boson popped into existence, even if it was for the most fleeting of moments (about 1022th of a second) by looking for a new particle with its own signature mass. The problem was, no one knew what the mass would be, so they didn't know where to look. It was a problem almost specifically designed for Big Data.

Finding the Higgs boson isn't so much like looking for a needle in a haystack as looking for a specific piece of hay in a haystack. Revealing the boson's existence was made all the more difficult because experimenters wouldn't be able to 'see' it directly. What they were looking for was the remnants of what it broke up into when it decayed into nothingness. The products of decay turned out to be two particles they already knew about. By recording the debris of trillions of subatomic particles, if they found a small bump in the population statistics at a specific energy level, the researchers at CERN could say they had found the Higgs boson.

This led to the creation of the giant atom smasher, located just outside Geneva in Switzerland. Formally, it is known as the Large Hadron Collider (LHC) and was built by CERN at a budgeted cost of €7.5bn when it was first switch on. In reality, it is definitely an atom smasher. It starts with a humble bottle of hydrogen, which is trickled into a chamber where the single electron that orbits the single proton in the hydrogen nucleus is stripped off, creating a stream of positively charged particles (hadrons). The protons then begin a process of acceleration. Imagine a series of Olympic hammer throwers whirling their hammers (a ball of hydrogen atoms on the end) round and round. The hydrogen atoms are passed from one hammer thrower to the next, each one increasing in size until the ball is split into two and given to two very large hammer throwers who are going in opposite directions. The balls of hydrogen are kept apart until they are made to collide inside detectors. It is the collision of these clumps of protons that creates the energy to reveal the subatomic particles.

At its widest, the LHC is 5.3 miles (8.5km) in diameter and the hydrogen atoms travel its 16.8-mile (27-km) circumference 11,245 times a second – close to the speed of light. The particles collide at a rate of 600 million collisions a second. The subatomic debris is captured by detectors at a rate of one petabyte of data per second. This raw data is filtered down into the most interesting events, but even when it is at full throttle, the LHC is pumping out 100 megabytes of data a second. By analyzing this data during the fleeting moments they exist, it was hoped that researchers could spot the hump in the data that revealed Higgs boson.

The LHC couldn't exist without Big Data and advanced computing. After all it was at CERN where Tim Berners-Lee and his colleagues invented the HTML language that gave rise to the World Wide Web in 1988. When it is in full operation, the LHC generates three gigabytes of data a second, and tens of petabytes a year. Like the genome project storage, data processing is a major headache. To solve this, right from the initial design it was always envisaged that data from the LHC would be processed and stored in a vast network of computers. By 2012, the Worldwide LHC Computing Grid distributed data to 170 computing centres in 36 countries. This giant super-computing facility has access to over half a million CPUs and 500 petabytes of storage. You can get involved, too. CERN has released some 300 terabytes of data through its Open Data Portal. CERN includes tools to make it easy for you to analyze the data, which can be downloaded to your computer.

Some people still worry that CERN is throwing away too much data. There is a genuine worry that the only thing the LHC will ever find is Higgs boson. If we don't find anything else, is this because we have reached the end of our physics, or are we just not looking in the right places? Some researchers say the data already exists for new discoveries; it's just that we haven't analyzed it to find the other particles such has been the concentration on the hunt for the Higgs boson. Dr Dan Hooper of the University of Chicago is more optimistic about the future of physics when he says the Higgs boson, 'Doesn't behave exactly as the Standard Model predicts'. He expects we are 'at the beginning of whole new era of discovery in particle physics'.

It is something of an understatement to say the confirmation of the Standard Model is one of mankind's greatest achievements, but there is further to go. China is planning a $6bn investment to create a super-collider twice the size of CERN, 50km in circumference, at Qinhuangdao in northern China. Construction could start in 2020. The intention is to create millions of Higgs bosons for study and maybe to throw up new exotic particles – the 'super partners' of the known particles, which could signal the explanation for dark matter, the stuff that holds galaxies together and stops stars flying off into the universe as galaxies rotate. It's the largest question in cosmology yet to be answered. Looking for dark matter is now part of an international effort to trawl through already existing data to see if we've overlooked the answer in the numbers we already have. Without Big Data and its filtering techniques, this wouldn't be possible.

It's tempting to ask what the everyday benefits of the LHC and the Higgs boson are to consumers and the people who pay for it – you and me. If it wasn't enough that we could understand the 'how' of why we are here, if not the 'why' of why we are here, then so far the project has laid the basis of the Internet, created the HTML language that still powers every website, created the largest computing grid in the world and shown that, given half the chance, we can, as a species, be united in a common cause when we really want to be. However, for those who need hard numbers to convince them of the benefits of basic research, a study of the socio-economic impact of the LHC found benefits from technology spill-over, supply-chain job creation, the training of today's scientists and the creation of an inspirational example to hold up to the next generation of researchers. Using a sophisticated modelling approach, it has been estimated there is a 90 per cent probability that the present value of the LHC project between 1993 and its expected decommissioning in 2025 will be €2.9 billion.

Maybe we should leave the final word to the man who, at first, foresaw little use for his discovery of an elemental particle:

> *In fact, research in applied science leads to reforms, research in pure science leads to revolutions, and revolutions, whether political or industrial, are exceedingly profitable things if you are on the winning side.* – J.J. Thompson, discoverer of the electron.

Big Data vs Science

We know how it ends. In about ten quadrillion years, all the stars will have burned out. What remains will be a cold, dark universe. Not to worry, though: long before this, the earth's molten metal core will have frozen and we will have plunged into the sun on the wildest roller-coaster ride ever seen. We even know how it began: about 14 billion years ago, a massive expansion of energy cooled in a fraction of a second and out of it, like raindrops from a mist, emerges the fundamental particles that form all of the matter of the known universe. Now, thanks to the Standard Model, we know pretty much everything about the 'in-between' bits with bewildering precision. As a species, it's a wonder we get out of bed in the morning any more.

There are gaps left in the Standard Model, but future generations of scientists will merely spend their time filling them in like an almost complete jigsaw. There is also the tiny problem of reconciling the Standard Model with Einstein's theory of general relativity, as the Standard Model doesn't explain how gravity comes about. It's a big problem, one which dogged Einstein himself in the later stages of his life. Reconciling the very big with the very small remains outside of our grasp's undoubtedly Big Data will play a role in finding the answer.

The technology surrounding genetic decoding has now been industrialised and the cost of a once monumentally expensive exercise is declining at a rate that exceeds Moores Law predictions, which assume that the power of a technology doubles every two years. The cost of gene sequencing collapsed in 2008 as second-generation technologies became available.

However, the morality and ethics of our genetic discovery can't be said to have kept pace with with the technology. Should we know this information just because we can? How is it going to be used? Already, life insurance forms contain questions asking whether you have had a genomic test and what the results were. Health companies might want higher premiums from those who have a high probability of becoming seriously ill in the future, if you haven't had them ironed out by CRISPR. Should you routinely screen your partner for regressive genes that make you both a bad match? It could have a major impact on our relationships and partner choices in the future. Meanwhile, the expectation is that personalised medicine could extend life expectancy by between ten and thirty years. Do we want to live that long without the quality of life that should come with it, but inevitably doesn't? Could we afford a society predominantly made up of people over a hundred years old?

Our scientific future is complex, exciting and on the cusp of a new age. There is no doubt that Big Data will help us sift through the possibilities during the remainder of the twenty-first century.

Man vs Big Data

There have been serious philosophical arguments put forward that suggest we may live in a vast computer-driven simulation. Philosophers Nick Bostrom and David Chalmers say that out of all the possibilities for humanity in the distant future, one of them has to be that we will become so advanced that we'll want to run simulations of our ancestry. Maybe we are living that simulation, which makes the universe the ultimate Big Data project. Bostrom and Chalmers believe this more likely than not but what is for sure is that the technologies and techniques falling under the term of "Big Data" are here to stay.

Big Data is mutating in front of us; gathering and storing vast amounts of data has become routine. The programs to analyze data through Machine Learning and Deep Learning can be plucked off the shelf and applied in seconds. It's become a seamless end to end process; a solution flailing around looking for problems.

So one of the first things we might need is a new term to supersede 'Big Data'. Undoubtedly, this rapid progress has led to some hype surrounding Big Data. It is well along the track of the rollercoaster in the classic lifecycle of new ideas: innovation trigger, peak of inflated expectations, trough of disillusionment, slope of enlightenment and plateau of productivity

– we are more than likely heading towards the peak of hysteria before the crash. The next stage, that will occupy the remainder of the century and beyond – will be the interesting part: enlightenment and the mature application of what we learned in the hype years.

One of the many challenges for Big Data may be that people become tired of being harvested for their data and they retreat behind the screen into the privacy of their own encrypted lives, away from cursor monitoring and predictions of their next move. As sure as something appears that it will

go on forever, humanity has a way, like water, of flowing around the obstacle. Big Data has to show itself to be a force for transformation and good rather than just another route to exploitation of the many by the few. Because what can be used for good can also be used for evil, and whatever the benefits of Big Data and its applications brings, it raises ethical questions that can be swept aside in the early flush of enthusiasm. When, eventually, a machine begins to show human-level intelligence and even experiences pain, would we have the right to shut it down? Is cancelling out that existence akin to taking a life? If a robot injures someone, who is to blame? The robot, the owner or the manufacturer? How will we have to amend or create existing laws to cope with the Fourth Industrial Revolution?

As a species, we appear to be ahead of the game in this respect. We are beginning to lay down the foundations of a mature information society for the near and not distant future. In an attempt to out run the developers, The European Parliament has begun the process of creating an ethical legal framework to ask not only what kind of things could we be doing in the new digital ocean, but also what kind of things should we be doing there. According to some, such as Luciano Floridi, professor of philosophy and information at the University of Oxford, 'we are stuck with the wrong conceptual framework … We need less science fiction and more philosophy'.

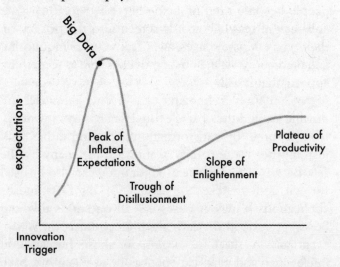

In the meantime, the benefits of Big Data are going to be many and tangible. New jobs, new products, new services and deeper insights into the world around us. At the 2017 O'Reilly Artificial Intelligence Conference in New York, leading-edge developers discussed AI applications using Big Data to;

- Create a device to count white blood cells in your home to detect cancer relapse early

- Design a program to detect the onset of sepsis before it takes hold

- Allow facial recognition technology to detect audience responses to television characters

- Detect driver emotions sending warnings and alerts to help prevent car accidents

- Create a collaborative system between AI and humans to write musical scores for advertisements and movies

- Design organic molecules and test their properties and solubility before going through the expense of manufacturing them

Big Data offers the availability of cheaper goods due to more efficient businesses, making experiences some people could only dream of a generation ago accessible to all. It holds out the possibility of personalised medicine and responsive, self-organizing cities and a world focused on prevention rather than cure. What's more, we are heading away from the old (deterministic) rule-based way of looking at things and heading towards probability as our guiding principle; it's a society based on managing risk like a gambler rather than expecting things to run like clockwork all of the time. Meanwhile, further advances in science (and maybe new branches of science) could be created from Big Data. It will refine our understanding of who we are and how we got here. And, yes, we may reach our ultimate destiny: your pet may yet receive a personalised greeting at your local pet store. It's going to be an exciting ride.

Acknowledgements

I would like to thank Lucy Warburton for persuading me to enter the world of Big Data and write this book. Melissa Smith has kept me on track when I might have disappeared into the backwaters of a subject that expands every day. My family have also had to put up with a lot; it must have been hard to listen to 'You'll never guess what I found out TODAY!!!', for a whole nine months. But most of all, this book is for our Cairn terrier, Toby. I SO want him to have his own personalised greeting when he enters a pet shop.

Sources

p.7 **All Brooke had said:** http://www.cbsnews.com/news/tv-news-anchors-report-accidentally-sets-off-viewers-amazons-echo-dots/

p.8 **Alexa uses voice:** http://uk.businessinsider.com/the-inside-story-of-how-amazon-created-echo-2016-4?r=US&IR=T

p.9 **Amazon's Alexa listens:** http://www.kpcb.com/blog/2016-internet-trends-report: slide 211

p.14 **...camera-based recording:** http://www.kpcb.com/blog/2016-internet-trends-report: slide 79

p.17 **For instance, the letter:** ASCII code table

p.23 **According to Mary Meeker's:** http://www.kpcb.com/blog/2016-internet-trends-report: slide 90

p.24 **In the two years:** http://www.kpcb.com/blog/2016-internet-trends-report: slide 78

p.27 **You name it:** http://bigdatasciencetraining.com/how-to-collect-big-data

On Twitter there are: http://fnuked.de/usaproxy/www2006-knowing-the-users-every-move--user-activity-tracking-for-website-usability-evaluation-and-implicit-interaction.pdf

Facebook's one billion: https://www.brandwatch.com/2016/05/47-facebook-statistics-2016/

Even more invasive: http://fnuked.de/usaproxy/www2006-knowing-the-users-every-move--user-activity-tracking-for-website-usability-evaluation-and-implicit-interaction.pdf

p.28 **Target, the large US:** http://www.ap-institute.com/big-data-articles/how-is-big-data-used-in-practice-10-use-cases-everyone-should-read.aspx

The film took: http://knowledge.ckgsb.edu.cn/2015/07/28/technology/the-power-of-big-data-in-china/

p.30 **This is a very:** Time to Period Double = ln(2)/ln(1 + r) where r is the growth rate expressed as a decimal.

p.31 **Amazon, Google, Facebook:** http://cloudtweaks.com/2015/03/surprising-facts-and-stats-about-the-big-data-industry/

Some people believe: http://www.independent.co.uk/environment/global-warming-data-centres-to-consume-three-times-as-much-energy-in-next-decade-experts-warn-a6830086.html

p.32 **In a 2015 survey:** http://www.kpcb.com/blog/2016-internet-trends-report: slide 210

p.33 **Fortunately, while the:** http://www.kpcb.com/blog/2016-internet-trends-report: slide 195

p.41 **Codd, who was:** http://www.seas.upenn.edu/~zives/03f/cis550/codd.pdf

In response to: https://en.wikipedia.org/wiki/Codd%27s_12_rules

p.45 **'So the user can...'** http://www.foxprohistory.org/interview_wayne_ratliff.htm

p.49 **Modern neural network:** https://en.wikipedia.org/wiki/Artificial_neural_network

p.52 **Since 2008:** http://www.kpcb.com/blog/2016-internet-trends-report: slide 122

p.55 **In 2016 an IBM:** https://futurism.com/artificially-intelligent-lawyer-ross-hired-first-official-law-firm/

The benign results: http://www.unglobalpulse.org/technology/hunchworks

p.56 **In one such:** https://en.wikipedia.org/wiki/Computer_Assisted_Passenger_Prescreening_System_II

For instance, there: http://iianalytics.com/research/15-ways-to-maximize-predictive-analytics-benefits-from-talent-assessments

p.59 **We will meet:** https://www.youtube.com/watch?v=vIkzWfftu4s

p.60 **Some hailed Snowden:** https://www.youtube.com/watch?v=tysIV6t54L4

In the US: https://en.wikipedia.org/wiki/Project_MINARET

p.61 **In 1972, legislation:** https://en.wikipedia.org/wiki/United_States_v._United_States_District_Court

George W. Bush assumed: https://en.wikipedia.org/wiki/Unitary_executive_theory

the 'Stellar Wind' project: https://en.wikipedia.org/wiki/Stellar_Wind

p.62 **His own recollection:** https://www.youtube.com/watch?v=vIkzWfftu4s&t=338s 'Citizenfour'

To help it along: https://fas.org/irp/budget

Even allowing for: https://www.bls.gov/data/inflation_calculator.htm - the inflation rate of 148% used in this calculation can be calculated here

Some $11bn of: http://www.washingtonpost.com/wp-srv/special/national/black-budget/ - I arrived at this estimate by using 15% proportion that went to the NSA in 2013 and applying it to the $72bn due to be spent in 2017.

p.67 (table) **Amazon:** https://www.amazon.co.uk/gp/help/customer/display.html/ref=footer_privacy?ie=UTF8&nodeId=502584

Cineworld: https://www.cineworld.co.uk/terms/privacy#privacy-policy

Facebook: https://www.facebook.com/about/privacy/

Jaguar Incontrol: https://incontrol.jaguar.com/jaguar-portal-owner-web/about/privacy-policy/GBR

LinkedIn: https://www.linkedin.com/legal/privacy-policy?trk=hb_ft_priv

Great Run Training: http://www.greatruntraining.org/privacy-policy/

Samsung: http://www.samsung.com/uk/info/privacy.html#none

Spotify: https://www.spotify.com/uk/legal/privacy-policy/#s7

Google: https://www.google.com/intl/en/policies/privacy/

p.68 **Our automated systems:** https://www.google.com/intl/en/policies/privacy/

p.69 **To address public:** https://www.commercce.gov/news/fact-sheets/2016/02/eu-us-privacy-shield

Specifically, Privacy Shield: http://ec.europa.eu/justice/data-protection/files/factsheets/factsheet_eu-us_privacy_shield_en.pdf

p.70 **The US, UK:** https://www.google.com/transparencyreport/userdatarequests/countries/?p=2015-12

Conversely, if you: Jaguar want £10 to provide me with a copy of what they hold on me. So does Great Run Training.

Please be aware: https://incontrol.jaguar.com/jaguar-portal-owner-web/about/privacy-policy/GBR

Meanwhile, Samsung has: http://www.samsung.com/uk/info/privacy.html#none

p.71 **... hack called 'Weeping Angel':** https://www.cnet.com/news/weeping-angel-hack-samsung-smart-tv-cia-wikileaks/

p.72 **Ross Ulbricht, for example:** https://en.wikipedia.org/wiki/Silk_Road_(marketplace)

However, his darkenet: https://www.theguardian.com/technology/2015/may/29/silk-road-ross-ulbricht-sentenced

In fact, The Tor: https://en.wikipedia.org/wiki/The_Tor_Project,_Inc

p.77 **Targets included Kim:** https://www.youtube.com/watch?v=hqKafI7Amd8

p.79 **My experience is:** http://www.spiegel.de/international/world/spiegel-interview-with-us-president-barack-obama-a-1122008.html

p.83 **Badoo, with 245:** https://en.wikipedia.org/wiki/Comparison_of_online_dating_websites

p.84 **Second comes:** https://en.wikipedia.org/wiki/Comparison_of_online_dating_websites

p.87 **There are a large number:** *Dataclysm: what our online lives tell us about our offline selves*, Christian Rudder (Fourth Estate, 2016).

p.90 **For two numbers:** The geometric mean is the Nth root of the scores where N is number of numbers being multiplied together. It's a mathematical trick to get around the problem of the more familiar arithmetic mean, which can be misleading, e.g. imagine a couple of 0% and 100% compatibility. If we used the arithmetic mean they would be (0 +100%)/2 = 50% compatible, which clearly isn't the case because one is obviously a saint and the other a psychopath. Using the geometric mean gets rid of this problem, i.e. $\sqrt{(0 \times 100\%)} = 0\%$

Match.com doesn't: http://www.slate.com/articles/news_and_politics/the_slate_quiz/2017/02/milo_yiannopoulos_uber_and_chris_christie_in_the_slate_news_quiz.html

p.91 **His girlfriend met:** http://www.slate.com/articles/life/ft/2011/07/inside_matchcom.html

p.92 **When Amy Webb:** 'How I hacked online dating' by Amy Webb: https://www.youtube.com/watch?v=d6wG_sAdP0U

The de-stigmatization of: https://www.forbes.com/sites/kevinmurnane/2016/03/02/pew-report-who-uses-mobile-dating-apps-and-online-dating-sites/#4355233666e3

Today, if you own: *Modern Romance: An Investigation* (p31), Ansari, Aziz and Klinenberg, Eric (Allen Lane, 2015).

However, the principle: http://metro.co.uk/2016/02/27/how-does-tinder-actually-work-5721632/

The company has at least: https://www.fastcompany.com/3054871/whats-your-tinder-score-inside-the-apps-internal-ranking-system

p.93 **According to an online:** *Dataclysm: what our online lives tell us about our offline selves* (p70), Christian Rudder (Fourth Estate, 2016).

p.94 **At worst, the conclusions:** *Dataclysm: what our online lives tell us about our offline selves*, Christian Rudder (Fourth Estate, 2016).

p.95 **It fills the 'eerie:** *Dataclysm: what our online lives tell us about our offline selves* (p109), Christian Rudder (Fourth Estate, 2016).

p.96 **Even if computer-based:** https://londonspermbankdonors.com

p.98 **Annalee Newitz, the:** Newitz, Annalee (31 August 2015). 'Ashley Madison Code Shows More Women, and More Bots'. *Gizmodo*. (Retrieved 19 December 2015)

... passed the Turing Test: Claire Brownell (11 September 2015). 'Inside Ashley Madison: Calls from crying spouses, fake profiles and the hack that changed everything'. *Financial Post*

p.103 **In reality, there are:** *The Looniness of the Long Distance Runner*, Russell Taylor (Andre Deutsch, 2003)

p.110 **Most commonly, it:** https://enduranceworks.net/blog/understanding-vo2-max-what-exactly-is-it-part-1/

VO2 max levels are generally: http://www.mytreadmilltrainer.com/vo2-max-men-women.html

p.112 **To find out where:** http://www.topendsports.com/testing/norms/vo2max.htm

p.114 **You certainly shouldn't:** *The Looniness of the Long Distance Runner*, Taylor

p.115 **Impair your ability:** https://runnersconnect.net/running-nutrition-articles/alcohol-and-endurance-running/

A bottle of wine: https://www.drinkaware.co.uk/alcohol-facts/alcoholic-drinks-units/alcohol-limits-unit-guidelines/

p.117 **Data from thousands:** http://journals.plos.org/ploscompbiol/article?id=10.1371/journal.pcbi.1000960

p.119 **Even allowing for:** Calculated by Stewart Cowley

The causes could: http://www.healthline.com/health-news/marathon-running-could-damage-kidneys#3

p.121**Better brain training:** http://www.runnersworld.com/racing/why-cant-i-run-faster

p.126 **between 2014 and:** http://www.gamblingcommission.gov.uk/Gambling-data-analysis/statistics/Industry-statistics.aspx

A 2016 report: http://www.gamblingcommission.gov.uk/Gambling-data-analysis/statistics/Industry-statistics.aspx

Still, the main: http://www.gamblingcommission.gov.uk/PDF/survey-data/Gambling-participation-in-2016-behaviour-awareness-and-attitudes.pdfs

p.129 **On 28 December 2011:** https://www.theguardian.com/sport/2012/mar/24/betfair-punters-voler-la-vedette

p.130 **It's important to:** https://en.wikipedia.org/wiki/Mathematics_of_bookmaking#Making_a_.27book.27_.28and_the_notion_of_overround.29

The odds of 'a to b': A relative probability of x represents odds of $(1 - x)/x$, e.g. 0.2 is $(1 - 0.2)/0.2 = 0.8/0.2 = 4/1$ (4-1, 4 to 1).

p.139 **Data on team:** http://www.sciencedirect.com/science/article/pii/S095070511300169X

Statisticians Håvard Rue: http://citeseerx.ist.psu.edu/viewdoc/download?doi=10.1.1.56.7448&rep=rep1&type=pdf

p.140 **Does a gambler:** By the way, I ran the latest odds through the EV calculator – River Plate fans need look away – they have simply no chance of winning, according to the bookies.

p.141**company called BETEGY:** http://money.cnn.com/2013/07/12/technology/innovation/betegy-soccer-betting/index.html?utm_source=feedburner&utm_medium=feed&utm_campaign=Feed%3A+rss%2Fmoney_latest+%28Latest+News%29

Database queries that: http://siliconangle.com/blog/2013/08/20/how-the-gambling-industry-is-betting-on-big-data

It beat four: *The Observer*, 5 February, 2017.

p.145 **When the quantity of any:** *The Wealth of Nations* (Chapter 7).

p.149 **As a student:** https://en.wikipedia.org/wiki/MONIAC

p.150 **For instance, a:** Chetty, Raj, John Friedman, and Jonah Rockoff. 2011. 'The Long-Term Impacts of Teachers: Teacher Value-Added and Student Outcomes in Adulthood.' NBER Working Paper no. 17699. Cambridge, MA: National Bureau of Economic Research.

Meanwhile, a study: Liran Einav, Dan Knoepfle, Jonathan Levin, and Neel Sundaresan. 2014. 'Sales Taxes and Internet Commerce.' *American Economic Review* 104(1): 1–26.

Sales fell by 80: *Made in America* (p402), Bill Bryson (Black Swan, 1998)

p.151 **Since online and offline:** http://www.mit.edu/%7Eafc/papers/Cavallo_Online_Offline.pdf

p.153 **while Twitter has been:** https://blog.twitter.com/2014/building-a-complete-tweet-index

p.157 **Well-known names:** http://www.zerohedge.com/news/2016-01-05/first-15-billion-hedge-fund-casualty-2016-blames-hfts-making-mockery-investing

p.158 **Suddenly, he was persona:** http://www.vanityfair.com/news/2010/04/fink-201004

It has 25 million: https://www.blackrock.com/aladdin/benefits/risk-managers

p.160 **In August 2015:** https://swissfinte.ch/wp-content/uploads/2016/06/FTPartnersResearch-DigitalWealthManagement.pdf

p.161 **During a briefing:** http://www.telegraph.co.uk/news/uknews/theroyalfamily/3386353/The-Queen-asks-why-no-one-saw-the-credit-crunch-coming.html

It took four: https://www.theguardian.com/uk/2012/dec/13/queen-financial-crisis-question

p.167 **What's more, the:** http://news.mit.edu/2015/harbinger-failure-consumers-unpopular-products-1223

p.170 **With almost freakish:** http://www.slideshare.net/mjft01/big-data-big-deal-a-big-data-101-presentation/16-Sexy_nerds_Data_ScientistThe_Sexiest

Demand for people: http://www.forbes.com/sites/louiscolumbus/2015/11/16/where-big-data-jobs-will-be-in-2016/#6b1760f6f7f1

p.171 **Thomas H. Davenport:** *Big Data at Work: Dispelling the Myths, Uncovering the Opportunities* (p88), Thomas H. Davenport (Havard Business Review Press, 2014)

p.174 **Some 72 per cent:** https://www.theguardian.com/science/2016/sep/01/how-algorithms-rule-our-working-lives

Saberr is a company: http://www.techworld.com/picture-gallery/startups/these-startups-will-help-you-hire-best-talent-or-find-your-dream-job-3637256/

p.175 **The resulting 'resonance:** http://www.techworld.com/startups/london-based-startup-saberr-can-predict-if-someone-is-right-fit-for-your-company-3632290/

Websites such as: https://klout.com/corp/score

It has been used: https://www.forbes.com/sites/erikkain/2012/04/25/do-you-have-klout-employers-want-to-know/#5230c81a110b

Since the 1970s: https://www.theguardian.com/women-in-leadership/2013/oct/14/blind-auditions-orchestras-gender-bias

p.177 the income share: https://obamawhitehouse.archives.gov/sites/whitehouse.gov/files/documents/Artificial-Intelligence-Automation-Economy.PDF

p.179 The US Bureau: https://obamawhitehouse.archives.gov/sites/whitehouse.gov/files/documents/Artificial-Intelligence-Automation-Economy.PDF

General Motors acquired: https://www.nytimes.com/2017/02/10/technology/ford-invests-billion-artificial-intelligence.html?imm_mid=0eda06&cmp=em-data-na-na-newsltr_20170222&_r=0

p.181 This compares to: *The Little Book of Hygge*, Meik Wiking (Penguin Life, 2016)

p.186 At the current rate: http://www.economist.com/news/essays/21596796-democracy-was-most-successful-political-idea-20th-century-why-has-it-run-trouble-and-what-can-be-do

In the 2016 US: http://heavy.com/news/2016/11/eligible-voter-turnout-for-2016-data-hillary-clinton-donald-trump-republican-democrat-popular-vote-registered-results/

p.187 Brexit, the UK's: https://ig.ft.com/sites/brexit-polling/

p.188 To add to the: http://www.usnews.com/news/the-report/articles/2015/09/28/why-public-opinion-polls-are-increasingly-inaccurate

p.190 Allowing for a: https://www.aaai.org/ocs/index.php/ICWSM/ICWSM10/paper/viewFile/1536/1842

p.192 Research by the University: https://en.wikipedia.org/wiki/Migration_in_China

p.193 In the French city: http://mashable.com/2013/09/25/big-data-cities/#B7zGxXZCgkqr

The New York Police: http://techpresident.com/news/24367/property-records-and-building-permits-part-new-nyc-open-data-release

p.193 Some social analysts: https://blogs.oii.ox.ac.uk/roughconsensus/2013/01/the-end-of-ideology-big-data-and-decision-making-in-politics/

p.194 A new trend: https://motherboard.vice.com/en_us/article/how-our-likes-helped-trump-win

p.195 Outside the West: http://www.economist.com/news/essays/21596796-democracy-was-most-successful-political-idea-20th-century-why-has-it-run-trouble-and-what-can-be-do

p.196 'There is in our: http://interglacial.com/pub/text/Umberto_Eco_-_Eternal_Fascism.html

p.202 The HGP was declared: https://en.wikipedia.org/wiki/Human_Genome_Project

The first person: https://en.wikipedia.org/wiki/Craig_Venter

p.203 As the price of: *Big Data, AI, the genome and everything*, Vijay Narayanan, March 2017

Comparing the storage: Big Data: *Astronomical or Genomical?*, Stevens ZD et al., PLOS: Biology July 2015

p.204 Two weeks later, Joshua: https://www.nytimes.com/2014/06/05/health/in-first-quick-dna-test-diagnoses-a-boys-illness.html?_r=0

p.205 Pac-Man molecule: https://www.broadinstitute.org/blog/machine-learning-approach-improves-crispr-cas9-guide-pairing

... up to 50 per cent: *Big Data, AI, the genome and everything*, Vijay Narayanan, March 2017

p.205 It may save: http://www.nhs.uk/news/2015/02February/Pages/Media-heralds-the-discovery-of-the-infidelity-gene.aspx

you only sequence: https://www.genome.gov/sequencingcosts/

p.208 DNA can survive: https://www.extremetech.com/extreme/134672-harvard-cracks-dna-storage-crams-700-terabytes-of-data-into-a-single-gram

The researchers were: https://en.wikipedia.org/wiki/DNA_digital_data_storage

p.209 'As a layman I think: https://www.theguardian.com/science/video/2012/jul/04/cern-higgs-boson-video

He lost $100: http://www.telegraph.co.uk/news/science/science-news/9376804/Higgs-boson-Prof-Stephen-Hawking-loses-100-bet.html

p.210 By recording the: *A Needle in the Haystack: Higgs Boson searches in the ATLAS Experiment*, Andrzej Olszewski, Marcin Wolter, Computing and Informatics, Vol. 32, 2013, 1256–1271

p.211 This giant super-computing: http://www.computerworld.com/article/2960642/cloud-storage/cerns-data-stores-soar-to-530m-gigabytes.html

CERN includes tools: http://www.ibtimes.co.uk/cern-releases-300tb-large-hadron-collider-data-so-everyone-can-study-particle-physics-free-1556597

p.212 Some people still: https://www.newscientist.com/article/mg21328564-700-is-the-lhc-throwing-away-too-much-data/

Dr Dan Hooper: The Higgs Boson – What we don't know, Dan Hooper. https://www.youtube.com/watch?v=sw4_9xhGzjo

Looking for dark: http://www.theleadsouthaustralia.com.au/industries/research-development/big-data-for-dark-matter/

p.213 Using a sophisticated: https://arxiv.org/abs/1603.00886: 'Forecasting the Socio-Economic Impact of the Large Hadron Collider: a Cost-Benefit Analysis to 2025 and Beyond', Massimo Florioi, Stefano Forte, Emanuela Sirtoi, March 2016.

In fact, research: *The Life of Sir J. J. Thomson* (p199), Rayleigh, John William Strutt (Cambridge University Press, 1943).

p.216 There have been: https://en.wikipedia.org/wiki/Simulation_hypothesis

p.217 ... peak of hysteria: Graphic from *Big Data and Transport: Understand and Assessing the Options*, 2015. Their source is Gartner Research

p.218 We need less: *Financial Times*, 23 February 2016, (p15). 'Opinion', Luciano Floridi.